Audel™

Installation Requirements of the 2002 National Electrical Code®
All New Edition

Audel™

Installation Requirements of the 2002 National Electrical Code® All New Edition

Paul Rosenberg

Wiley Publishing, Inc.

Vice President and Executive Group Publisher: Richard Swadley
Vice President and Executive Publisher: Robert Ipsen
Vice President and Publisher: Joseph B. Wikert
Executive Editorial Director: Mary Bednarek
Editorial Manager: Kathryn A. Malm
Executive Editor: Carol Long
Senior Production Editor: Fred Bernardi
Development Editor: Regina Brooks
Production Editor: Pamela M. Hanley
Text Design & Composition: TechBooks

For general information on our other products and services please contact our Customer Care Department within the United States at (800) 762-2974, from outside the United States at (317) 572-3993, or by fax at (317) 572-4002.

Library of Congress Cataloging-in-Publication Data:
CIP data to come

ISBN: 0-764-54278-8

Printed in the United States of America

10 9 8 7 6 5 4 3 2 1

Contents

Introduction

Very few people who have had to work with the *National Electric Code®(NEC)* would argue with the statement that it is a very difficult, complex, and confusing document. Not many professional electricians have a thorough knowledge of it; much less the homeowner who is trying to upgrade his or her wiring system properly. The purpose of this book is to arrange all of the pertinent requirements (and *only* the pertinent requirements) of the *National Electric Code* in a manner that is "user-friendly"; allowing the reader to find the needed information painlessly and very quickly.

The problem with the *NEC* is that many communities use it as law, and as such, it must be written accordingly. Every possible facet of every of every type of electrical installation must be covered. Because of this, each section of the *NEC* is full of design criteria, engineering requirements, installation requirements, and a host of exceptions—all in engineering lingo and "legalese." It's not hard to see why it is such a difficult document.

In order to help understand and apply the *NEC*, a great number of guides have been written, most of which have a legitimate place. These guides serve to make all parts of the code understandable. They are written for engineers, designers, installers, and inspectors.

The book you now hold in your hands is substantially different from standard code guidebooks. Rather than seeking to write a new guidebook covering everything in the *NEC*, we have eliminated all of the purely engineering and design regulations, and concentrated only on the requirements for electrical installations. By omitting the engineering and design requirements, most of the confusion of the *NEC* is eliminated in one stroke. This leaves only the rules that matter when actually installing the electrical wiring—which is the only reason the code is referred to, 99% of the time.

In addition to doing away with all of the *NEC* requirements that do not pertain to installing electrical wiring, the entire text of the book has been arranged according to the type of installation. This way, the reader can simply turn to the appropriate section for the type of wiring being installed, and quickly find all of the installation requirements for the work at hand. Aside from the requirements mentioned for each section, no other requirements should apply to the work covered therein. All of the many electricians who have reviewed this arrangement of installation requirements have been enthusiastic about it.

This book is designed exclusively for the installer of electrical wiring, and is the result of many years supervising and instructing electricians in the requirements of the *NEC*. Every effort has been made to make this book as easy to use as possible, both for the professional electrician and for the homeowner who wishes to do his or her own electrical work safely and efficiently, and perhaps avoid hassles with the local electrical inspector.

For actually installing electrical wiring this book should be far more useful than the standard Code Handbooks. For engineering questions, however, the *National Electrical Code* should be consulted.

I trust that you will find this arrangement of installation requirements as useful as others have. A few projects should save you more than enough time and money to cover its expense.

Finally, please remember that good workmanship and safety-consciousness are essential ingredients for any good electrical installation. Like fire, electricity can be the best of friends or the worst of foes. Without careful workmanship and an overriding concern for the safety of the installation and the installer, no electrical installation is worthwhile.

Paul Rosenberg

Chapter 1

General Requirements

General Requirements (*Article 110*)

The first requirement of the *National Electrical Code* is that all installations must be performed "in a neat and workmanlike manner." In other words, all electrical installation requirements presuppose an installer who is concerned, informed, and thinking. Without this prerequisite, any other requirements are almost worthless.

Basic Requirements (*Article 110*)

The following basic requirements apply to all electrical installations. They should be reviewed periodically by every electrical installer.

All unused openings in boxes, cabinets, etc. must be filled.

All equipment must be securely mounted. Wooden plugs in masonry are not allowed.

Panelboards and other exposed buswork must be protected from paint, plaster, or other similar materials during the construction process.

Conductors in manholes must be racked to provide a reasonable amount of access space.

Free circulation of air around electrical equipment, especially equipment that requires such air flow for sufficient heat removal, can't be obstructed.

All electrical connections *must* be made with devices that are listed and clearly marked as suitable for the intended use.

Conductors must be spliced with suitable splicing devices, or by soldering, brazing, or welding. Soldered splices must first be joined, so that the splice is not dependent on the solder for mechanical or electrical strength. All splices must be covered with insulation equivalent to that of the conductors.

Wire connectors (lugs, wire nuts, etc.) must be rated no lower than the operating temperature of the conductors they are used with.

A reasonable amount of working space must be provided around all electrical equipment. Generally, the minimum is 3 feet. *Table 110.26(A)(1)* shows specific requirements.

Lighting, enough to work on the equipment, must be provided in the areas around electrical equipment.

Except for panels of 200 amps or less in dwelling units, there must be a minimum head space in front of electrical panels of 7 feet.

All live parts operating at over 50 volts must be guarded against accidental contact by persons or objects. See *Section 110.31* for operation at over 600 volts. The primary methods of accomplishing this are as follows:

Locating the equipment in a room that is accessible only to qualified persons.

Installing permanent and effective partitions or screens.

Locating the equipment on a balcony or platform that will exclude unqualified persons.

Locating the equipment 8 feet or more from the floor.

Guards must be installed to protect electrical equipment from physical damage where necessary.

Entrances to rooms or areas where there are live parts must have a sign posted forbidding unqualified persons from entering.

All disconnecting means (service, feeder, or branch circuit) must be marked, showing the purpose. This is not required if the purpose of the disconnecting means is obvious.

Exposed parts of high- and medium-voltage systems must have adequate clearance above working spaces. *Table 110.34(E)* lists these distances.

Use of Grounded Conductors (*Article 200*)

All premises wiring systems must have a grounded conductor, except where the NEC specifically permits otherwise.

A grounded conductor must have insulation that is equal to that of any ungrounded conductors it is used with.

A grounded premises wiring system must receive its power from a grounded supply system.

A grounded conductor No. 6 or smaller must be covered with white or natural gray insulation (Figure 1.1), except:

1. Fixture wires.

2. Aerial cables can use a ridge on the grounded conductor, rather than a different color of insulation.

3. Where only qualified persons will have access to the conductors, colored conductors can be taped or painted white or gray at their terminations.

4. Grounded conductors in MI cables can be identified otherwise.

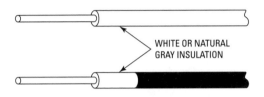

NO. 6 OR SMALLER WIRE SHALL HAVE
WHITE OR NATURAL GRAY INSULATION

WIRE LARGER THAN NO. 6 MAY HAVE
WHITE TAPE OR PAINT TO INDICATE
GROUNDED CONDUCTORS

Figure 1.1 Method of identifying grounded conductors.

Grounded conductors No. 4 or larger can be identified either by having white or gray insulation, by having three longitudinal stripes 120 degrees apart, or by having a white marking at terminations.

If grounded conductors of different systems are installed in common boxes, raceways, etc., the first system must be marked as above, the second system's grounded conductor must be identified by having white insulation with a colored (but *not* green) tracer, and any other systems must have their own means of identification.

Cables to switches can use the grounded (white) conductor to bring power to the switch, but not *from* the switch (Figure 1.2).

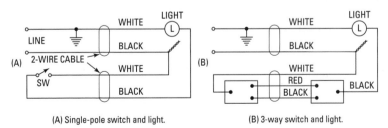

(A) Single-pole switch and light.   (B) 3-way switch and light.

Figure 1.2 Method of connecting a common light and switch.

Terminals used specifically for grounding conductors must be identified by a color sufficiently different from that used for other terminals.

For devices with screwshells, the grounded conductor must be connected to the screwshell, not to the tab, as shown in Figure 1.3.

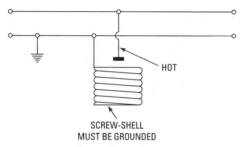

Figure 1.3   Grounding of a screwshell base.

Chapter 2

Branch Circuits

Branch Circuits (*Article 210*)

Classifications

Branch circuits are rated based on the setting of their overcurrent protection devices. Branch circuits to individual items can have any amperage rating, but branch circuits feeding multiple items must be of one of the following classifications: 15, 20, 30, 40, or 50 amperes.

Multi-outlet circuits greater than 50 amperes can be used only where they will be accessible only to qualified persons, and not for lighting circuits.

Multi-wire circuits can be considered branch circuits as long as all conductors originate from the same panelboard (Figure 2.1).

Multi-wire branch circuits are allowed to supply line only to neutral loads, except if the circuit supplies only one piece of equipment, or if all ungrounded conductors of the circuit open simultaneously by the same branch-circuit overcurrent device.

In dwellings, a multi-wire branch circuit that supplies more than one wiring device on the same yoke must have all ungrounded conductors of the circuit open simultaneously by the same branch-circuit overcurrent device.

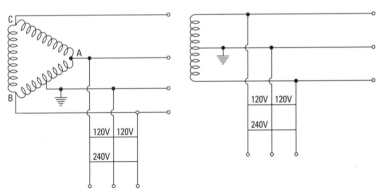

Figure 2.1   (A) One type of multi-wire circuit from a 4-wire delta system. (B) A 3-wire, 120/240 V, multi-wire circuit.

If more than one voltage system is used in a building, the systems must be separately identified and marked at each panel. Usually this is done with the ungrounded-conductor insulation: black, red, blue for 120/208 V systems; brown, orange, yellow for 277/480 V systems; etc.

Grounded Conductors

The grounded circuit conductor (usually called the neutral conductor) of branch circuits must be white or natural gray.

If grounded conductors of different systems are installed in common boxes, raceways, etc., the first system must be marked as above, the second system's grounded conductor must be identified by having white insulation with a colored (but *not* green) tracer, and any other systems must have their own means of identification.

The equipment-grounding conductor of a branch circuit must be either green, or green with a yellow tracer.

Voltages

Voltages greater than 120 volts between conductors are not allowed in dwellings, hotels, or motels at the following outlet locations:

> At lampholders.
>
> At cord-and-plug–connected appliances 1440 watts or less, or less than ¼ horsepower.

Circuits with voltages between conductors greater than 120 volts are allowed to supply the following:

> Medium-base screwshell or other types of lampholders.
>
> Electric discharge lighting auxiliary equipment (ballasts).
>
> Cord-and-plug or permanently connected utilization equipment.

Circuits between 120 and 277 volts to ground are allowed to supply the following:

> Electric discharge lighting fixtures that have medium-base screwshell lampholders.
>
> Other types of lampholders that have appropriate voltage ratings.
>
> Lighting fixtures that have mogul-base screwshell lampholders.
>
> Electric discharge lighting auxiliary equipment (ballasts).
>
> Cord-and-plug or permanently connected utilization equipment.

Circuits between 277 volts to ground and 600 volts between conductors are allowed to supply the following:

> Electric discharge lighting auxiliary equipment (ballasts) in permanently mounted fixtures. The fixtures must also be at least 22 feet high, when mounted on poles; or 18 feet high, when mounted on buildings.
>
> Cord-and-plug or permanently connected utilization equipment.

Receptacles and Cord Connectors

Receptacles installed on 15- or 20-amp circuits must be of the grounding type, except to replace receptacles where no grounding means exists. See Figure 2.2 for grounding extensions of ungrounded circuits.

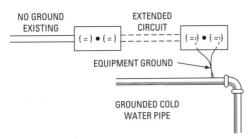

Figure 2.2   Grounding equipment on an existing circuit that is not grounded.

Grounding contacts of receptacles must be solidly grounded, except where ground-fault interrupter (GFI) receptacles replace ungrounded receptacles in locations where no grounding means exists, or for vehicle-mounted equipment.

When existing receptacles are replaced in areas that require ground-fault receptacles for new installations (bathrooms, near sinks, etc.), they must be replaced with GFI receptacles or have GFI breakers installed for the circuit(s) involved.

The grounding mentioned above must be done by connection to the equipment-grounding conductor.

In locations where more than one type of system (ac and dc, different voltages, different frequencies, etc.) is available, all receptacles must be noninterchangeable, so that a plug for one system won't fit into the receptacle of another system.

Ground-Fault Circuit Interrupters

Only ground-fault circuit interrupter–protected 15- and 20-amp receptacles may be installed in the following locations in dwellings:

Bathrooms.

Accessible parts of garages.

Outdoors, where there is access to the receptacle directly from grade (6 feet from ground or less).

In crawl spaces at or below grade level.

In unfinished basements (receptacles for sump pumps, laundry circuits, or cord-and-plug–connected single appliances, etc.).

Countertop receptacles within 6 feet of a kitchen sink or wet bar sink.

In boathouses.

All 15- and 20-amp receptacles in hotel and motel bathrooms must have ground-fault protection.

Circuits derived from autotransformers must have ground-fault protection, except for autotransformer output circuits that have a grounded conductor directly connected to the grounded conductor that supplies the autotransformer; or for an autotransformer used to boost an existing 208-volt circuit to 240 volts, and having a grounded output conductor as mentioned above.

Circuits with no grounded conductors are allowed to be taped from circuits with grounded conductors. Switching devices in the taped circuits must have a pole in each ungrounded conductor. If multipole switches function as a disconnecting means, all conductors must open simultaneously when the device is activated.

Arc-Fault Circuit Interrupters

An arc-fault circuit interrupter (AFCI) protects persons and equipment from an arc fault by recognizing the characteristics unique to an arc fault and deenergizing the circuit when an arc fault is detected. All branch circuits supplying 15 or 20 A, single-phase, 125 V outlets installed in dwelling unit bedrooms must be AFCI protected by a listed device that protects the entire branch circuit. This change extends AFCI protection to all 125 V outlets in dwelling unit bedrooms—outlets containing receptacles or otherwise. The 1999 NEC required AFCI protection only for all branch circuits that supply 15 or 20A, single-phase, 125 V receptacles in dwelling unit bedrooms. The Code defines an outlet as a point on the wiring system at which current is taken to supply utilization equipment. This includes openings for receptacles, luminaires, or smoke detectors.

The traditional practice of separating the lighting from the receptacle circuits in dwelling unit bedrooms will, with the addition of this requirement, require two AFCI circuit breakers. The 125 V limitation to the requirement means AFCI protection is not required for a 240 V circuit, such as one for an electric heater.

Branch-Circuit Ratings

Branch-circuit conductors can't have an ampacity less than the load connected to them.

Branch-circuit conductors that supply ranges, wall-mounted ovens, cooktops, or other household cooking appliances must have a rating at least as high as the branch-circuit rating and the connected load.

Ranges rated at over 8¾ kW must have a branch circuit rated at least 40 amperes.

Tap conductors are allowed from a 50-ampere circuit to a range, wall-mounted oven, cooktop, or other household cooking appliance. They must run no longer than necessary, and must be rated at least 20 amperes.

The neutral conductor of a circuit that supplies a range, wall-mounted oven, cooktop, or other household cooking appliance can be smaller than the ungrounded conductors of the circuit. The size must be based on *Table 220.19*, column A, can't be smaller than No. 10 copper, and must be rated at least 70% of the circuit rating.

Tap conductors are allowed from a 40- or 50-ampere circuit to loads other than household cooking appliances. They must run no longer than necessary, and must be rated at least 20 amperes.

Tap conductors are allowed from a circuit under 40 amperes to loads other than household cooking appliances. They must run no longer than necessary, and must be rated at least 15 amperes.

Branch-circuit conductors and equipment must be protected by overcurrent protective devices.

Heavy-duty lampholders must be used for circuits rated over 20 amperes.

A single receptacle installed on an individual branch circuit must have a rating at least that of the branch circuit.

For branch circuits that supply more than one outlet, the maximum cord-and-plug–connected load can't be more than that shown in *Table 210.21(B)(2)*.

Receptacle ratings for various sizes of circuits are shown in *Table 210.21(B)(3)*. Either 15- or 20-amp rated receptacles are allowed on 20-amp circuits, and only 15-amp receptacles are allowed on 15-amp circuits.

The rating of any cord-and-plug–connected appliance can't be greater than 80% of the circuit's rating.

A summary of branch-circuit requirements is given in *Table 210.24*.

Appliance Outlets

Appliance outlets, including laundry outlets, must be installed within 6 feet of the appliance.

Receptacles in dwellings (garages, basements, bathrooms, and hallways are excluded) must be installed so that no point along the wall space is more than 6 feet from a receptacle outlet. This also includes wall spaces more than 2 feet in width. Sliding panels in exterior walls don't have to be considered as wall space. Floor receptacle outlets that are close to a wall can be counted as if they were wall receptacles (Figure 2.3).

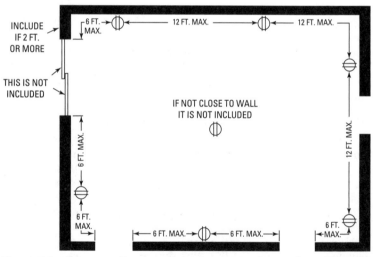

Figure 2.3   Proper wall receptacle outlet spacing for residential housing.

In addition to the requirement above, two or more 20-amp receptacle outlets are required for small appliances in a dwelling unit's kitchen, dining room, or breakfast room. Outlets for dishwashers, disposals, etc. don't count. No other loads are to be connected to these circuits except electric clocks or outdoor receptacles.

At least two receptacle circuits are required in every kitchen. The two receptacle outlets mentioned above can be used to fill this requirement, if desired.

In kitchen and dining areas of dwellings, every counter space 1 foot wide or greater must have a receptacle outlet. These receptacles must be placed so that no countertop space is more than 24 inches from a receptacle outlet. All countertop receptacles within 6 feet of the sink must be ground-fault protected (Figure 2.4).

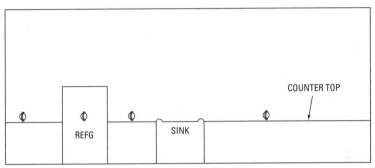

Figure 2.4   Proper wall receptacle outlet spacing for countertops in kitchen or dining areas.

Island or peninsula countertops 12 inches by 24 inches or greater must have receptacles installed so that no point on the perimeter of the countertop is more than 24 inches from a receptacle. These receptacles must be mounted above the countertop, or within 12 inches below the countertop. Note that no receptacles are allowed to be installed face up in countertops.

At least one receptacle must be installed in each bathroom, adjacent to the basin location (or adjacent to any sink or basin that is used for human grooming). This receptacle must be ground-fault protected.

At least two outdoor receptacles, one near the front entrance and one near the rear entrance, must be installed for all dwelling units (single-family or two-family). Two outdoor receptacles are required for single-family units; four must be installed for a two-family dwelling unit (two for each unit). These receptacles must be ground-fault protected.

Dwelling units must have at least one receptacle outlet for a laundry area. This is not required for apartment buildings that have a common laundry area, or for multi-family dwellings where laundry facilities are not permitted.

In addition to the above requirement, single-family dwellings must have at least one receptacle outlet in each basement, attached garage, or separate garage with electric power. These receptacles must be ground-fault protected.

Receptacle outlets are required for hallways more than 10 feet long.

Guest rooms in hotels and motels must conform to the receptacle requirements given above. However, hotels are given more leeway for receptacle spacing requirements to allow for their furniture spacing.

Show windows must have receptacles for every 12 inches of window space, measured horizontally.

Except for rooftop equipment for one- and two-family dwellings, receptacles must be installed for the servicing of heating, air conditioning, and refrigeration equipment. The receptacle must be on the same level as the equipment, within 25 feet. It can't be connected to the load side of the equipment's disconnecting means. Except when used in single- or two-family residences, all such roof-mounted receptacles must be ground-fault protected.

At least one wall switch–controlled lighting outlet must be installed in each habitable room. Switched wall outlets can be considered lighting outlets for this requirement, except for kitchens.

In addition to the above, switched lighting outlets are also required at the entry to an attic, crawl space, utility room, or basement.

Branch circuits in dwelling units can supply only equipment in that unit, or loads associated only with that dwelling unit. All circuits such as central alarm systems, common area lighting, etc. must come from a "house panel," and not from a dwelling unit panel. This applies even when the dwelling unit is occupied by a building manager.

Chapter 3

Feeders

Feeders (*Article 215*)

Ratings and Sizes

Feeder conductors must have an ampacity no less than required to carry their load.

Feeders can be no smaller than 30 amperes when the load being supplied consists of the following types of circuits:

Two or more branch circuits supplied by a 2-wire feeder.

More than two 2-wire branch circuits supplied by a 3-wire feeder.

Two or more 3-wire branch circuits supplied by a 3-wire feeder.

Two or more 4-wire branch circuits supplied by a 3-phase 4-wire feeder.

Feeders must be protected with overcurrent protection devices.

Grounding and Other Requirements

Feeders that contain a common neutral are allowed to supply two or three sets of 3-wire feeders, or two sets of 4- or 5-wire feeders.

In any metal raceway or enclosure, all feeders with a common neutral must have their conductors run together.

When a feeder supplies branch circuits that require equipment grounding conductors, the feeder must supply a grounding means, to which the equipment grounding conductors (from the branch circuits) can be connected.

Two-wire ac or dc circuits of two or more ungrounded conductors are allowed to be taped from ungrounded conductors of circuits that have a grounded neutral. Switching devices in each taped circuit must have a pole in each ungrounded conductor.

For a 4-wire delta system, the phase conductor having the higher voltage must be identified by an orange marking.

Feeders that supply 15- or 20-ampere branch circuits that require ground-fault protection can be protected against ground faults, rather than protecting individual circuits or receptacles.

Outdoor Feeders or Branch Circuits

When open individual conductors (for circuits 600 volts or less) are run in overhead spans, the following sizes must be used:

> For spans 50 feet or less—No. 10 copper, No. 8 aluminum, or larger.
>
> Spans over 50 feet—No. 8 copper, No. 6 aluminum, or larger.

Overhead conductors for festoon lighting must be no smaller than No. 12, unless supported by a messenger wire.

Circuits between 120 volts between conductors and 277 volts to ground can supply lighting fixtures that illuminate outdoor areas, or commercial or public buildings. The fixtures must be at least 3 feet away from windows, fire escapes, etc.

Wiring 600 volts or under is allowed on buildings. The wiring must be in one of the following:

> Rigid metal conduit.
>
> Intermediate metal conduit.
>
> Electrical metallic tubing.
>
> Rigid nonmetallic conduit.
>
> As open conductors or messenger-supported wiring.
>
> Multiconductor cables.
>
> MC or MI cables.
>
> Cable trays.
>
> Cablebus.
>
> Wireways.
>
> Auxiliary gutters.
>
> Flexible metal conduit.
>
> Liquid-tight flexible metal conduit.
>
> Liquid-tight flexible nonmetallic conduit.
>
> Busways.

Outdoor circuits must enter exterior walls with an upward slant, so that water will tend to flow away from the interior of the building.

Open conductors not over 600 volts must have the following clearances over grade:

> Above finished grade, sidewalks, platforms, etc. from which the conductors could be reached by pedestrians (not from

vehicles); and where the voltage is not more than 150 volts to ground—10 feet.

Over residential driveways and commercial areas not subject to truck traffic; and where the voltage is not more than 300 volts to ground—12 feet.

For areas in the above classification (12-foot rating) when the voltage is greater than 300 volts to ground—15 feet.

Over public streets, alleys, roads, parking areas subject to truck traffic, driveways on nonresidential property, and other land traversed by vehicles (orchards, grazing, etc.)—18 feet.

Feeder or branch conductors can't be installed where they will obstruct, or be installed underneath, openings in farm and commercial buildings through which materials can be moved.

Conductors not over 600 volts must have an 8-foot clearance over roofs. This clearance must be maintained within 3 feet of the roof surface, measured horizontally.

If a roof is subject to pedestrian traffic, it is considered the same as a sidewalk.

If a roof has a slope of 4 inches rise for every 12 inches run or greater, the clearance can be only 3 feet. The voltage can't be more than 300 volts between conductors.

If no more than 4 feet of conductors pass over a roof overhang, and they are terminated by a through-the-roof raceway or approved support, and the voltage is not more than 300 volts between conductors, only 18 inches clearance is required.

Horizontal clearance from signs, chimneys, antennas, etc. need be only 3 feet.

When these conductors attach to a building, they must be at least 3 feet from windows, fire escapes, etc.

Buildings three stories or 50 feet in height must have horizontal spaces left between overhead conductor runs 6 feet wide, to allow for fire-fighting ladders.

Vegetation (trees) can't be used to support overhead conductor spans, except for temporary wiring.

Chapter 4

Services

Services (*Article 230*)

Service Conductors

Service conductors are not allowed to pass through a building or structure, and then supply another building or structure, unless they are encased in 2 inches or more of concrete.

Conductors are not considered in a building (although they actually are) in any of the following circumstances:

> If they are encased in 2 inches or more of concrete.
>
> If they are enclosed in a raceway, and then enclosed in 2 inches of brick.
>
> If they are in proper transformer vaults.

The only conductors outside of service conductors that are allowed in service raceways are grounding conductors, and load management conductors that have overload protection.

When a service raceway enters from underground, it must be sealed to prevent the entry of gas. Empty raceways must also be sealed.

Service cables must be kept at least 3 feet from windows or similar openings, but are allowed with less clearance over windows (rather than next to them).

Service conductors must have enough ampacity to carry the load placed on them.

Service conductors are allowed to be no smaller than No. 8 copper or No. 6 aluminum. When services feed only limited loads of single branch circuits, No. 10 copper (this is equivalent to No. 12 hard-drawn, which the Code actually specifies) can be used.

The size of the neutral conductor for a service must be at least as follows:

> Service conductors 1100 kcmil or smaller—the neutral must be at least as large as the grounding electrode conductor indicated in *Table 250.66*.
>
> Service conductors larger than 1100 kcmil—the neutral must be at least 12.5% of the size of the largest phase conductor.

For parallel phase, the neutral must be 12.5% of the equivalent cross-sectional conductor area.

Each service can have only one set of service conductors, except that multiple tenants or occupants of one building can have separate service entrance conductors.

One set of service conductors is allowed to supply a group of service entrance enclosures.

Service entrance conductors must be of sufficient size to carry their load.

Service conductors can't be spliced, except as follows:

1. Clamped or bolted connections in meter fittings are allowed.

2. Service conductors can be taped to supply two to six disconnecting means grouped at a common location.

3. At a proper junction point where the service changes from underground to overhead.

4. A connection is allowed when service conductors are extended from an existing service drop to a new meter, and then brought back to connect to the service entrance conductors of an existing location.

5. Sections of busway are allowed to be connected together to build the service.

Service Clearances

Service conductors not over 600 volts must have the following clearances over grade (Figure 4.1):

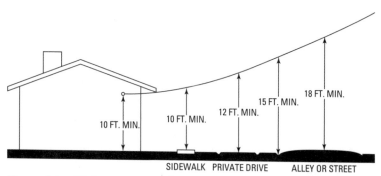

Figure 4.1   Minimum service-drop clearance.

Above finished grade, sidewalks, platforms, etc. from which the conductors could be reached by pedestrians (not from vehicles); and where the voltage is not more than 150 volts to ground—10 feet.

Over residential driveways and commercial areas not subject to truck traffic, and where the voltage is not more than 300 volts to ground—12 feet.

For areas in the above classification (12-foot rating) where the voltage is greater than 300 volts to ground—15 feet.

Over public streets, alleys, roads, parking areas subject to truck traffic, driveways on nonresidential property, and other land traversed by vehicles (orchards, grazing, etc.)—18 feet.

Conductors not over 600 volts must have an 8-foot clearance over roofs. This clearance must be maintained within 3 feet of the roof surface, measured horizontally. The requirement for maintaining 3 feet of clearance beyond the edge of the roof does not apply in situations where the conductors are attached to the side of the structure.

If a roof is subject to pedestrian traffic, it is considered the same as a sidewalk.

If a roof has a slope of 4 inches rise for every 12 inches run or greater, the clearance can be only 3 feet. The voltage can't be more than 300 volts between conductors.

If no more than 4 feet of conductors pass over a roof overhang and are terminated by a through-the-roof raceway or approved support, and the voltage is not more than 300 volts between conductors, only 18 inches clearance is required (Figure 4.2).

Horizontal clearance from signs, chimneys, antennas, etc. need be only 3 feet.

When these conductors attach to a building, they must be at least 3 feet from windows, fire escapes, etc.

The point of attachment of service conductors to a building must be no lower than the above-mentioned clearances, but never less than 10 feet.

Where service masts are used to support service drops, they must be of sufficient strength or be supported for braces or guys.

Underground Service Conductors

Underground service conductors must be suitable for the conditions existing where they are installed. They must be protected where required.

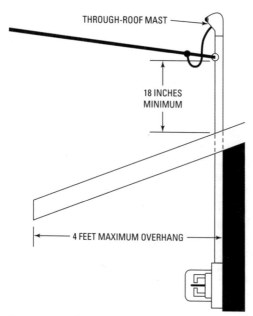

THROUGH-ROOF MAST

18 INCHES
MINIMUM

4 FEET MAXIMUM OVERHANG

Figure 4.2   A service-drop mast mounted through the roof.

Bare service grounding conductors are allowed as follows:

Bare copper conductors in a raceway.

Bare copper conductors directly buried, when soil conditions are suitable.

Bare copper conductors directly buried, without regard to soil conditions, when installed as part of a cable assembly approved for direct burial.

Bare aluminum conductors directly buried, without regard to soil conditions, when installed as part of a cable assembly approved for installation in a raceway or direct burial.

Wiring Methods

Service conductors 600 volts or less can be installed using any of the following methods:

Rigid metal conduit.

Intermediate metal conduit.

Electrical metallic tubing.

Electrical nonmetallic tubing (ENT).

Service entrance cables.

Wireways.

Busways.

Cablebus.

Open wiring on insulators.

Auxiliary gutters.

Rigid nonmetallic conduit.

Type MC cable.

Type IGS cable.

Mineral-insulated, metal-sheathed cable.

Liquid-tight flexible nonmetallic conduit.

Flexible metal conduit, but only for runs of 6 feet or less, between raceways or between raceways and service equipment. An equipment bonding jumper must be run with the conduit.

Cable tray systems are allowed to support service conductors.

Service entrance cables installed near sidewalks, driveways, or similar locations must be protected by one of the following methods:

Rigid metal conduit.

Intermediate metal conduit.

Rigid nonmetallic conduit, when suitable for the location.

Electrical metallic tubing.

Other approved methods.

Service entrance cables must be supported within 12 inches of every service head, gooseneck, or connection to a raceway or enclosure. They must be supported at intervals no longer than 30 inches.

Individual open conductors must be mounted on insulators or insulating supports, and supported as indicated in *Table 230.51(C)*.

Cables that are not allowed to be installed in contact with buildings must be mounted on insulators or insulating supports, and must be supported every 15 feet or less. A cable must be supported in such a way that there will be no less than 2 inches clearance between the cable and the covered surfaces.

Services must enter exterior walls with an upward slant, so that water will tend to flow away from the interior of the building. Drip loops must be made.

Service raceways that are exposed to the weather must be rain-tight and arranged so that they will drain.

Service raceways must have a rain-tight service head where they connect to service drops.

Service cables, unless they are continuous from a pole to the service equipment, must be provided with a service head or shaped into a gooseneck. When shaped into a gooseneck, the cable must be taped and painted, or taped with self-sealing, weather-resistant thermoplastic.

Except where it is not practical, service heads for service entrance cables must be higher than the point of attachment of the service-drop conductors.

Drip loops must be made below the level of the service head or the end of the cable sheath.

Service raceways and cables must terminate in boxes or enclosures that enclose all live parts.

For a 4-wire delta system, the phase conductor having the higher voltage must be identified by an orange marking.

Service Equipment

Energized parts of service equipment must be enclosed and guarded.

A reasonable amount of working space must be provided around all electrical equipment. Generally, the minimum is 3 feet. *Table 110.16(A)* lists specific requirements.

Service equipment must be suitable for the amount of short-circuit current that is available for the specific installation.

A means must be provided for disconnecting all service entrance conductors from all other conductors in a building. A terminal bar is sufficient for the neutral conductor.

The service-disconnecting means must be installed in an accessible location outside the building, or, if the disconnecting means is to be installed inside the building, at the point nearest to where the service conductors enter the building.

All service disconnects must be suitable for the prevailing conditions.

The disconnecting means for each service can't consist of more than six switches or circuit breakers mounted in a single enclosure.

Individual circuit breakers controlling multi-wire circuits must be linked together with handle ties.

Multiple disconnects must be grouped and marked to indicate the load being served.

Additional service disconnects for emergency power, stand-by systems, fire pumps, etc. can be separated from other service equipment.

Each occupant in a building must have access to his or her own service-disconnecting means.

A service-disconnecting means must simultaneously open all ungrounded conductors.

The service-disconnecting means can be either power operable or manual. When power operable, a manual override in case of a power failure must be possible.

The disconnecting means must have a rating no less than the load being carried.

Service-disconnecting means for single-family dwellings must have a minimum rating of 100 amperes; service disconnects for all other occupancies must have a rating of at least 60 amps.

Smaller service sizes are permissible for limited loads, not in occupancies. For loads of two 2-wire circuits, No. 8 copper or No. 6 aluminum conductors can be used. For loads with one 2-wire circuit, No. 12 copper or No. 10 aluminum conductors can be used. These may never be smaller than the branch-circuit conductors.

Only the following items are permitted to be connected to the line side of service disconnects:

Cable limiters or current-limiting devices.

Meters operating at no more than 600 volts.

Disconnecting means mounted in a pedestal and connected in series with the service conductors, located away from the building being supplied.

Instrument transformers (current or potential transformers).

Surge protection devices.

High-impedance devices.

Load management devices.

Taps that supply load management devices, circuits for emergency systems, fire pump equipment, stand-by power equipment, and fire and sprinkler alarms that are provided with the service equipment.

Solar photovoltaic or other interconnected power systems.

Control circuits for power-operable disconnects. These must be provided with their own overcurrent protective and disconnecting means.

Ground-fault protection devices that are part of listed equipment. These must be provided with their own overcurrent protective and disconnecting means.

When more than one building or structure on the same property is under single management, each structure must be provided with its own service-disconnecting means (Figure 4.3).

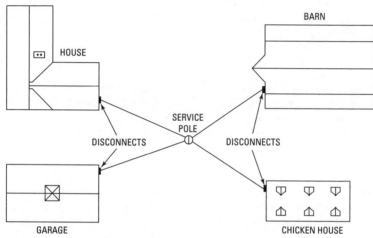

Figure 4.3 A group of farm buildings with a service drop to each.

Overcurrent Protection

An overcurrent protection device must be installed in each conductor and must be rated no higher than the ampacity of the service conductor.

No overcurrent device is allowed in a grounded conductor, as it would represent a safety hazard.

Services of 1000 amps or more for solidly grounded wye systems must have ground-fault protection.

Each occupant in a building must have access to his or her own service overcurrent protection devices.

Where necessary to prevent tampering, an automatic overcurrent protection device supplying only a single load can be locked.

The overcurrent protective means must cover all equipment except the following:

1. Cable limiters or current-limiting devices.

2. Meters operating at no more than 600 volts.

3. Instrument transformers (current or potential transformers).

4. Surge protection devices.

5. High-impedance shunts.

6. Load management devices.

7. Taps that supply load management devices, circuits for emergency systems, fire pump equipment, stand-by power equipment, and fire and sprinkler alarms that are provided with the service equipment.

8. Solar photovoltaic or other interconnected power systems.

9. Control circuits for power-operable disconnects. These must be provided with their own overcurrent protection and disconnecting means.

10. Ground-fault protection devices that are part of listed equipment. These must be provided with their own overcurrent protection and disconnecting means.

11. The service-disconnecting means.

Services Over 600 Volts

Service conductors over 600 volts can be installed using any of the following methods:

Rigid metal conduit.

Intermediate metal conduit.

Electrical metallic tubing (EMT).

Busways.

Cablebus.

Open wiring on insulators.

Rigid nonmetallic conduit.

Suitable metal-clad cables.

MV cables, bare conductors, and bare busbars are allowed in areas that are accessible only to qualified persons.

Cable tray systems can support service conductors.

Conductors and supports must be of sufficient strength to withstand short-circuits.

Open wires must be guarded.

Cable conductors emerging from a metal sheath or raceway must be protected by potheads.

Chapter 5

Overcurrent Protection

Overcurrent Protection (*Article 240*)

All conductors (except for some flexible cords) must have overcurrent protection no greater than their ampacity.

There are modifications to the above requirements for specific types of circuits (some motor circuits, some remote-control circuits, etc.) that are covered by their specific articles.

The standard ampere ratings for fuses and inverse-time circuit breakers are as follows:

15	80	250	1000
20	90	300	1200
25	100	350	1600
30	110	400	2000
35	125	450	2500
40	150	500	3000
50	175	600	4000
60	200	700	5000
70	225	800	6000

Fuses are also rated 1, 3, 6, 10, and 601 amps.

For the purpose of calculations, adjustable trip circuit breakers are considered rated at their maximum possible setting.

Fuses or circuit breakers can't be connected in parallel unless factory assembled in parallel.

Supplementary protection, such as in-line fuses installed in lighting fixtures, can't replace branch-circuit overcurrent protection.

No overcurrent device can be connected in series with any grounded conductor, except if the overcurrent device opens all conductors of the device, so that no conductor can operate independently.

When a change in size is made in the ungrounded conductor, a similar change can be made in the grounded conductor.

Unless specifically excepted, overcurrent devices must be accessible.

Overcurrent devices must be connected at the point where the circuit to be protected receives its supply.

Taps

The requirement that all overcurrent devices must be connected at the point where the circuit receives its supply is excepted for taps. The requirements for taps are as follows.

If taps are to be made from feeder conductors, they must (Figure 5.1):

Be no longer than 10 feet.

Terminate in an overcurrent protection device.

Have an ampacity no less than the device being fed or the overcurrent device mentioned above.

Extend no further than the device they supply.

Be enclosed in a raceway.

Have an ampacity at least 10% that of the overcurrent protection device on the line side.

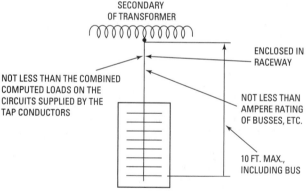

Figure 5.1 A tap circuit not to exceed 10 feet in length.

Feeder taps up to 25 feet long can be made under the following conditions (Figure 5.2):

The tap must terminate in a branch-circuit protective device.

The tap must have an ampacity of at least one-third of the feeder ampacity. For taps made to transformer secondaries, the secondary must have an ampacity that, when multiplied by the secondary-to-primary voltage ratio, is at least

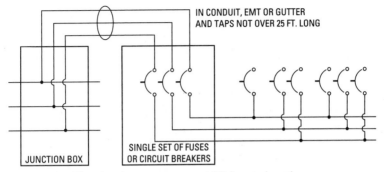

Figure 5.2   Tap circuits not to exceed 25 feet in length.

one-third of the ampacity of the conductors or overcurrent device from which the primary conductors are tapped.

The tap must be protected from physical damage.

In high-bay manufacturing buildings (which are more than 35 feet from floor to ceiling, measured at the walls), taps longer than 25 feet are permitted. In these cases:

The tap conductors must have an ampacity of at least one-third that of the feeder conductors.

The tap conductors must terminate in an appropriate circuit breaker or set of fuses.

The tap conductors must be protected from damage and installed in a raceway.

The tap conductors must be continuous, with no splices.

The minimum size of tap conductors is No. 6 AWG copper or No. 4 AWG aluminum.

The tap conductors can't penetrate floors, walls, or ceilings.

The tap conductors may be run no more than 25 feet horizontally, and no more than 100 feet overall.

Tap conductors are allowed from a 50-ampere circuit to a range, wall-mounted oven, cooktop, or other household cooking appliance. They must run no longer than necessary, and must be rated at least 20 amps.

Circuits with no grounded conductors are allowed to be tapped from circuits with grounded conductors. Switching devices in the tapped circuits must have a pole in each ungrounded conductor. If multipole switches function as a disconnecting means, all conductors must open simultaneously when the device is activated.

Tap conductors are allowed from a 40- or 50-ampere circuit to loads other than household cooking appliances. They must run no longer than necessary, and must be rated at least 20 amperes.

Tap conductors are allowed from a circuit under 40 amperes to loads other than household cooking appliances. They must run no longer than necessary, and must be rated at least 15 amperes.

All branch-circuit or feeder taps from a busway must use plug-ins or connectors that have overcurrent devices in them. This is not required in only two situations:

> Where fixtures that have an overcurrent device mounted on them are mounted directly onto the busway.

> Where an overcurrent device is part of a cord plug for cord-connected fixed or semi-fixed lighting fixtures.

For group motor installations, taps to single motors don't need branch-circuit protection in any of the following cases:

> The conductors to the motor have an ampacity equal to or greater than that of the branch-circuit conductors.

> The conductors to the motor are no longer than 25 feet, they are protected, and the conductors have an ampacity at least one-third as great as the branch-circuit conductors.

Secondary conductors can be tapped from transformers of separately derived systems, when the following conditions are met:

> The conductors are not longer than 25 feet.

> All overcurrent devices are grouped.

> The tap conductors are protected.

> The ampacity of the conductors is no less than the secondary current rating of the transformer, and the sum of the overcurrent devices will limit this capacity.

Locations

Overcurrent devices must be located where they won't be subject to physical damage.

Overcurrent devices should be enclosed in cabinets or enclosures and mounted in the vertical position unless impractical.

When installed in damp or wet locations, overcurrent devices must be mounted with at least a ¼-inch air space between the enclosure and the surface on which they are mounted (Figure 5.3).

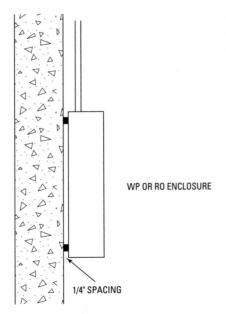

Figure 5.3 Spacing of enclosures for overcurrent devices in damp or wet locations.

WP OR RO ENCLOSURE

1/4" SPACING

Fuses and Circuit Breakers

Edison-base fuses are for replacement only.

A disconnecting means must be installed on the line side of fuses operating at more than 150 volts to ground. Disconnect switches are built this way.

Circuit breakers to be used as switches must be marked SWD.

Chapter 6

Grounding

Grounding (*Article 250*)
Circuit and System Grounding

Two-wire dc systems that supply wiring in a premises must be grounded, unless one of the following situations exists:

The system supplies only industrial equipment in limited areas, and has a ground detector.

The system operates at 50 volts or less between conductors.

The system operates at over 300 volts between conductors.

The system is taken from a rectifier, and the ac supplying the rectifier is from a properly grounded system.

The system is a dc fire-protective signaling circuit that has a maximum current of 0.03 amperes.

The neutral conductor of all 3-wire dc systems must be grounded.

Alternating-current circuits operating at less than 50 volts must be grounded if any of the following conditions exist:

The circuit is installed as overhead wiring outside of buildings.

The circuit is supplied by a transformer, and the transformer supply circuit is grounded.

The circuit is supplied by a transformer, and the transformer supply circuit operates at over 150 volts to ground.

Alternating-current systems operating at between 50 and 1000 volts, and supplying premises wiring, must be grounded if any of the following conditions exist:

The system can be grounded so that the voltage to ground of ungrounded conductors can't exceed 150 volts.

The system is a 3-phase, 4-wire wye, and the neutral is used as a circuit conductor.

The system is a 3-phase, 4-wire delta, and the midpoint of one phase is used as a circuit conductor.

When a grounded service conductor is not insulated the following situations are excepted from this requirement:

Systems used only to supply industrial electric furnaces used for melting and refining.

Separately derived systems used only to supply rectifiers that supply adjustable-speed drives.

Separately derived systems from transformers with 1000 V or less primaries and used only for controls, or where only qualified persons will service the installation, or where continuity of control power is required, or where ground detectors are installed on the control system, or where high-impedance neutral systems are installed according to *Section 250.36.*

Other specialized systems (see *Section 250.20[B]*, exceptions).

Refer to the diagrams of Figure 6.1.

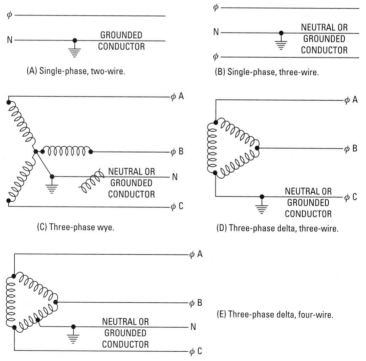

Figure 6.1 Grounding different types of circuits.

Circuits for cranes operating in class III locations over combustible fibers may *not* be grounded.

Location of Grounding Conductors

If the use of multiple grounding paths creates a problem (in certain circumstances it can), one of the following steps can be taken:

Discontinue one or more of the grounding locations.

Change the locations of the connections.

Discontinue one or more (but not all) of the objectionable ground currents.

Employ other remedies that are approved by the authority having jurisdiction (the local inspector).

The above rule should not be taken as allowance for not having every item connected to the system grounded.

Direct-current systems that have to be grounded must have the grounding connection made at one or more supply stations. These connections can't be made at individual services or on premises wiring. If the dc source is located on the premises, a grounding connection can be made at the first disconnecting means or overcurrent device.

AC System Grounding Connections

AC systems that require grounding must have a grounding electrode conductor at each service that connects to a grounding electrode. The grounding electrode conductor must be connected to the grounded service conductor at an accessible location between the load end of the service and the grounding terminal (Figure 6.2).

Where the transformer supplying the load is located outside the building, a separate connection must be made between the grounded service conductor and a grounding electrode, either at the transformer or at another location *outside* the building.

No grounding connection (to the grounding electrode) can be made on the load side of the service-disconnecting means. If this is done, it sometimes results in problematic and potentially dangerous ground-current loops.

If an ac system that operates at less than 1000 volts is grounded, the grounded conductor must be run to every service-disconnecting means and bonded to every disconnecting means enclosure. This conductor must be run with the phase conductors and can't be smaller than the grounding electrode conductor (specified in

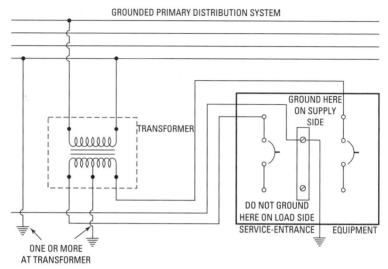

Figure 6.2 Grounding on the supply side of the service-entrance equipment will provide a ground on the supply if it is ever disconnected.

Table 250.66). If the phase conductors are larger than 1100-kcmil copper or 1750-kcmil aluminum, the grounded conductor must be at least 12.5% of the area of the largest phase conductor. If the service conductors are paralleled, the size of the grounded conductor must be based on the total cross-sectional area of the largest set of phase conductors.

When more than one service-disconnecting means is located in a listed assembly, only one grounded conductor must be run to and bonded to the service enclosure.

Two or More Buildings Supplied by a Common Service

If two or more buildings are supplied by a single service, the grounded system in each building must have its own grounding electrode, connected to the building disconnecting means enclosure. The grounding electrode of each building must have a connection to the grounded service conductor on the load side of the service-disconnecting means.

A grounding electrode is not required in separate buildings where only one branch circuit is present and does not require grounding.

Where two or more buildings are supplied by an ungrounded service, each structure must have a grounding electrode that is connected to the metal enclosure of the building or the structure's disconnecting means. The grounding electrode is not required in separate buildings where only one branch circuit is present and does not require grounding.

Disconnecting Means Located in Separate Building on Same Premises

When one or more buildings are under the same management and the disconnects are remotely located, the following conditions must be met:

> The neutral is to be connected to the grounding electrode at the first building only.
>
> Any building with two or more branch circuits has to have a grounding electrode. The equipment-grounding conductor from the first building must be run to this building with the phase conductors, and connected to the grounding electrode just mentioned.
>
> The connection of the grounding conductor to the grounding electrode must be made in a junction box enclosure or cabinet located just inside or just outside the building.
>
> If the grounding conductor is run in metal conduit or cable, it must be bonded to the conduit or cable at both ends.
>
> The grounding conductor, if run underground, must be insulated if livestock are present.

Grounding conductors must be sized according to *Table 250.122.*

Conductor to Be Grounded

The following conductors must be grounded (refer to Figure 6.1):

> One conductor of a single-phase, 2-wire system must be grounded.
>
> The neutral conductor of a single-phase, 3-wire system must be grounded.
>
> The center tap of a wye system that is common to all phases must be grounded.
>
> A delta system must have one phase grounded.
>
> In a delta system where one phase is used as shown above, the mid-tap must be grounded and must be used as the neutral conductor.

Grounding for Separately Derived Systems

When separately derived systems must be grounded, the following requirements must be met:

> Bonding jumpers must be used to connect the equipment-grounding conductors of the derived system to the grounded conductor. This connection can be made anywhere between the service-disconnecting means and the source, or it can be made at the source of the separately derived system, if it has no over-current devices or disconnecting means. The bonding jumper can't be smaller than the grounding electrode conductor (specified in *Table 250.66*). If the phase conductors are larger than 1100-kcmil copper or 1750-kcmil aluminum, the bonding jumper must be at least 12.5% of the area of the largest phase conductor. If the service conductors are paralleled, the size of the bonding jumper must be based on the total cross-sectional area of the largest set of phase conductors.
>
> A grounding electrode conductor must be used to connect the grounded conductor of the derived system to the grounding electrode. This connection can be made anywhere between the service-disconnecting means and the source, or it can be made at the source of the separately derived system, if it has no overcurrent devices or disconnecting means. The grounding electrode conductor must be sized as specified in *Table 250.66*. If the phase conductors are larger than 1100-kcmil copper or 1750-kcmil aluminum, the bonding jumper must be at least 12.5% of the area of the largest phase conductor. If the service conductors are paralleled, the size of the bonding jumper must be based on the total cross-sectional area of the largest set of phase conductors. Class I circuits that are derived from transformers rated no more than 1000 VA don't require a grounding electrode if the system grounded conductor is bonded to the grounded transformer case.
>
> A grounding electrode must be installed next to (or as close as possible to) the grounding electrode. The grounding electrode can be any of the following:
>
>> The nearest structural metal part of the structure that is grounded.
>>
>> The nearest metal underground water pipe that is grounded.
>
> An electrode, usually No. 4 (minimum) bare copper wire or reinforcing bar, at least 20 feet long and in direct contact with

the earth for its entire length. If reinforcing bar is used, it must be at least ½ inch in diameter.

A ring of No. 2 (minimum) bare copper wire, at least 2½ feet below grade and encircling the structure (at least 20 feet long).

A ground rod, plate electrode, or made electrode.

High-Impedance Grounded Neutral Connections

Where high-impedance neutral systems are allowed (see *Section 250.36*), the following requirements must be observed:

The grounding impedance must be installed between the grounding electrode and the neutral conductor.

The neutral conductor must be fully insulated and must have an ampacity that is no less than the maximum current rating of the grounding impedance. The neutral may not, however, be smaller than No. 8 copper or No. 6 aluminum.

The system neutral can have no other connection to ground, except through the impedance.

The neutral conductor between the system source and the grounding impedance can be run in a separate raceway.

The connection between the equipment grounding conductors and the grounding impedance (properly called the equipment bonding jumper) must be unspliced from the first system-disconnecting means to the grounding side of the impedance.

The grounding electrode conductor can be connected to the grounded conductor at any point between the grounding impedance and the equipment-grounding connections.

Grounding of Enclosures

All metal enclosures must be grounded, except:

1. Metal boxes for conductors that are added to knob-and-tube, open wiring, or non–metal-sheathed cable systems that have no equipment grounding conductor. The runs can be no longer than 25 feet and must be guarded against contact with any grounded materials.

2. Short runs of metal enclosures used to protect cable runs.

Equipment Grounding

All exposed non–current-carrying metal parts of equipment that are likely to become energized must be grounded if any of the following conditions exist:

>If the equipment is within 8 feet vertically, or 5 feet horizontally, of ground or any grounded metal surface that can be contacted by persons.
>
>If the equipment is located in a damp or wet location, unless isolated.
>
>If the equipment is in contact with other metal.
>
>If the parts are in hazardous locations.
>
>When the enclosure is supplied by a wiring method that supplies an equipment ground, such as metal raceway, metal-clad cable, or metal-sheathed cable.
>
>When equipment operates at over 150 volts to ground, except:
>
>>**1.** Non-service switches or enclosures that are accessible only to qualified persons.
>>
>>**2.** Insulated electrical heater frames, by special permission only.
>>
>>**3.** Transformers (and other distribution equipment) mounted over 8 feet above ground or grade.
>>
>>**4.** Listed double-insulated equipment.

The following types of equipment, regardless of voltage, *must* be grounded:

>All switchboards, except insulated 2-wire dc switchboards.
>
>Generator and motor frames of pipe organs, except if the generator is insulated from its motor and ground.
>
>Motor frames.
>
>Motor controller enclosures, except for ungrounded portable equipment or lined covers of snap switches.
>
>Elevators and cranes.
>
>Electric equipment in garages, theaters, and motion picture studios, except pendant lampholders operating at 150 volts or less.
>
>Electric signs.
>
>Motion picture projection equipment.
>
>Equipment supplied by class I, II, III, or fire-protective signaling circuits, except where specifically cited otherwise.

Motor-operated water pumps.

The metal parts of cranes, elevators and elevator cars, mobile homes and recreational vehicles, and metal partitions around equipment above 100 volts between conductors.

Except when specifically allowed otherwise, all cord-and-plug–connected equipment must be grounded.

Metal raceways, frames, raceways, and other non–current-carrying parts of electrical equipment must be kept at least 6 feet away from lightning protection conductors and lightning rods.

Methods of Grounding

Equipment grounding connections must be made as follows:

For grounded systems—by bonding the equipment-grounding conductor and grounded service conductor to the grounding electrode conductor (Figure 6.3).

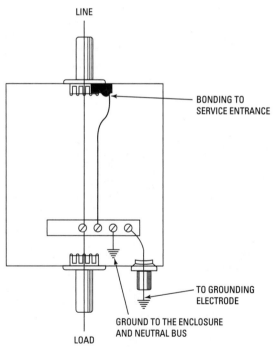

Figure 6.3 The grounding conductor that grounds the neutral must also be used for all other grounding.

For ungrounded systems—by bonding the equipment-grounding conductor to the grounding electrode conductor.

Grounding receptacles that replace nongrounded receptacles can be bonded to a grounded water pipe.

The conductor path to ground from equipment and metal enclosures must:

Be permanent and continuous.

Have enough capacity for any fault current imposed on it.

Have a low enough impedance so as not to limit the voltage to ground.

The earth can't be used as the only grounding conductor. In other words, you can't ground something by simply attaching a wire to its case and pushing the wire into the ground.

Only one grounding electrode is allowed at each building. Two or more grounding electrodes that are bonded together are considered the same as one grounding electrode.

Metal sheaths of underground service cables can be considered grounded, only because of their contact with the earth and bonding to the underground system. They don't need to be connected to the grounding electrode conductor or grounding electrode. This is also the case if the cable is installed underground in metal conduit and bonded to the underground system.

Non–current-carrying metal parts of equipment, raceways, etc. that require grounding can meet this requirement by being connected to an appropriate equipment-grounding conductor.

Electrical equipment is considered grounded if it is secured to grounded metal racks or structures designed for the support of the equipment. Mounting equipment on the metal frame of a building is *not* considered sufficient for grounding.

See Chapter 31 for the specific grounding requirements of mobile homes.

Except where specifically permitted, the neutral conductor is *never* to be used to ground equipment on the load side of the service disconnect.

If a piece of equipment is connected to more than one electrical system, it must have an appropriate ground connection for each system.

Bonding

The following parts of service equipment must be bonded together (Figure 6.4):

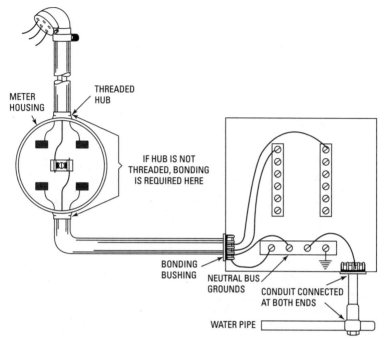

Figure 6.4 An example of proper bonding.

Raceways, cable trays, and cable armor or sheaths.

Enclosures, meter fittings, etc.

Raceway or armor enclosing a grounding electrode conductor.

One exposed means must be provided for the bonding of other systems (such as telephone or cable TV systems). This can be done by one of the following means:

Grounded metal raceway (must be exposed).

An exposed grounding electrode conductor.

Any other approved means, such as extending a separate grounding conductor from the service enclosure to a terminal strip.

The various components of service equipment must be bonded together by one of the following methods:

Bonding the components to the grounded service conductor.

Connection with rigid or intermediate metal conduit, made with threaded couplings, or threadless couplings and connectors that are made up tight. Standard locknut and bushing connections are *not* sufficient.

Connecting bonding jumpers between the various items.

Using grounding locknuts or bushings connected to an equipment-grounding conductor running between the various items.

Refer also to Figure 6.5.

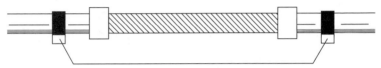

Figure 6.5 Bonding must extend around any flexible conduit used in conjunction with service-entrance equipment.

Metal raceways and metal-sheathed cables that contain circuits operating at more than 250 volts to ground must be bonded by the same means as for service equipment, shown immediately above, except for connection to the grounded service conductor. If knock-outs in enclosures or boxes are field cut, standard methods can be used.

If metal service cable has an uninsulated neutral in direct contact with the metal armor or sheath, the sheath or armor is considered grounded without any further connection.

A receptacle's grounding terminal must be bonded to a metal box in which it is installed by one of the following means (Figure 6.6):

Connection with a jumper wire, commonly called a pigtail.

Receptacle yokes and screws that are approved for this purpose.

Direct metal-to-metal contact between the box and the receptacle's yoke. This is for surface-mounted boxes *only*.

Installation in floor boxes that are listed for the purpose.

Connection to an isolated grounding system, where required to eliminate electrical noise in sensitive circuits.

All electrical components that are allowed to act as equipment conductors must be well fitted together to ensure electrical

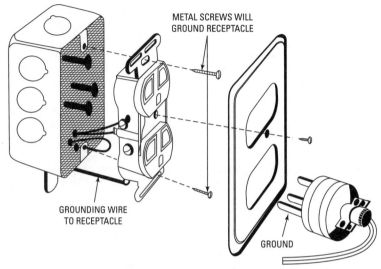

METAL SCREWS WILL
GROUND RECEPTACLE

GROUNDING WIRE
TO RECEPTACLE

GROUND

Figure 6.6 A grounding jumper shall be used from grounded boxes to the grounding terminal of the receptacle.

continuity. They *must* be well bonded. These systems include the following:

> Metal raceways.
> Cable trays.
> Cable armor.
> Cable sheaths.
> Enclosures.
> Frames.
> Fittings.

All metal raceways must be made electrically continuous. This requires special care at expansion joints, etc.

Main bonding jumpers connect the grounded service conductor and the service enclosure. This connection should be made according to the instructions supplied by the manufacturer of the service equipment. The bonding jumper can't be smaller than the grounding electrode conductor (specified in *Table 250.66*). If the phase conductors are larger than 1100-kcmil copper or 1750-kcmil aluminum, the bonding jumper must be at least 12.5% of the area of the largest phase conductor. If the service conductors are paralleled, the size of the bonding jumper must be based on the total cross-sectional area of the largest set of phase conductors.

An equipment-bonding jumper on the load side of the service must be sized according to *Table 250.122*, based on the largest overcurrent that protects the conductors in the raceways or enclosures.

The bonding jumpers mentioned above can be inside or outside of the equipment being bonded. If installed outside, the jumper can be no more than 6 feet long and must be routed along with the equipment.

Interior metal water piping systems *must* be bonded to one of the following:

> The service equipment enclosure.
>
> The grounded conductor, at the service only.
>
> The grounding electrode conductor, unless it is too small.
>
> The grounding electrode.

The bonding jumper mentioned above *must* be sized according to *Table 250.66*.

Other metal piping systems that may be energized must be bonded. The size of the jumper must be based on *Table 250.122*, using the rating of the circuit likely to energize the piping for calculations.

Exposed steel building frames that are not intentionally grounded must be bonded to either the service enclosure, the neutral bus, the grounding electrode conductor, or a grounding electrode. The bonding conductor must be sized according to *Table 250.66*.

Grounding Electrode Systems

All of the items mentioned below are suitable for grounding electrodes. All of the following (where available) must be bonded together, forming the grounding electrode system:

> The nearest structural metal part of the structure that is grounded.
>
> The nearest metal underground water pipe that is grounded.
>
> An electrode, usually No. 4 (minimum) bare copper wire or reinforcing bar, at least 20 feet long and in direct contact with the earth for its entire length. If reinforcing bar is used, it must be at least ½ inch in diameter, and it must be either bare or

have some type of electrically conductive coating, such as the zinc that is used for galvanization.

A ring of No. 2 (minimum) bare copper wire, at least 2½ feet below grade and encircling the structure (at least 20 feet long).

A ground rod, plate electrode, or made electrode.

Gas piping or aluminum electrodes are not acceptable.

Interior metal water piping is not allowed to be used as a grounding electrode conductor.

Grounding Conductors

A grounding electrode conductor can be copper, aluminum, or copper-clad aluminum. Its size must be determined according to *Table 250.66*, or if the phase conductors are larger than 1100-kcmil copper or 1750-kcmil aluminum, the grounding electrode conductor must be at least 12.5% of the area of the largest phase conductor. If the service conductors are paralleled, the size of the grounding electrode conductor must be based on the total cross-sectional area of the largest set of phase conductors.

The grounding electrode conductor can't be spliced. Bus bars and required taps are excepted.

An equipment-grounding conductor can be any of the following, and must be sized according to *Table 250.122*:

A copper (or other type that is corrosion resistant) conductor.

Rigid metal conduit.

Intermediate metal conduit.

Electrical metallic tubing.

Listed flexible metal conduit, in lengths of 6 feet or less.

The armor of type AC cable.

The sheath of mineral-insulated, metal-sheathed cable.

The sheath of type MC cable.

Cable trays.

Cablebus.

Other metal raceways that are electrically continuous.

Six feet is the maximum length of any combination of flexible metal conduit, flexible metallic tubing, and liquid-tight flexible metal conduit that can be used in any grounding run.

Grounding electrode conductors can be installed in any of the following:

Rigid metal conduit.

Intermediate metal conduit.

Rigid nonmetallic conduit.

Electrical metallic tubing.

Cable armor.

Grounding electrode conductors No. 6 copper or larger can be attached directly to a building surface; they are not required to be in a raceway.

Isolated metal parts of outline lighting systems can be bonded together by a No. 14 copper or No. 12 aluminum conductor that is physically protected.

No automatic cutout or switch can be placed in the grounding conductor.

Bare or insulated copper or aluminum grounding electrode conductors are forbidden in direct contact with masonry or earth, or where subject to corrosive conditions.

Grounding Conductor Connections

The connection between a grounding electrode conductor and a grounding electrode must be accessible, permanent, and effective. The safety of the entire system often depends on this connection.

Where metal piping systems are used as grounding electrodes, all insulated joints or parts that can be removed for service must have jumpers installed around them. The jumpers must be the same size as the grounding electrode conductor (Figure 6.7).

All grounding connections must be made by listed means, and must be permanent and secure (Figure 6.8).

If grounding connections are made in areas where they can be subjected to physical damage, they must be protected.

If more than one grounding conductor is present in a box or enclosure, all grounding conductors in the box must be connected. This connection must be made in such a way that the removal of any one device, etc. can't affect the connection. In common language, grounding connections can't feed through; they must pigtail (Figure 6.9).

A metal box must be connected to a grounding conductor, whether it be a conduit used as the grounding conductor or a separate grounding wire.

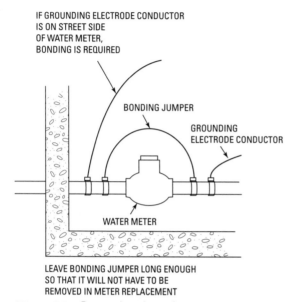

IF GROUNDING ELECTRODE CONDUCTOR
IS ON STREET SIDE
OF WATER METER,
BONDING IS REQUIRED

BONDING JUMPER

GROUNDING
ELECTRODE CONDUCTOR

WATER METER

LEAVE BONDING JUMPER LONG ENOUGH
SO THAT IT WILL NOT HAVE TO BE
REMOVED IN METER REPLACEMENT

Figure 6.7 Proper bonding of a water meter.

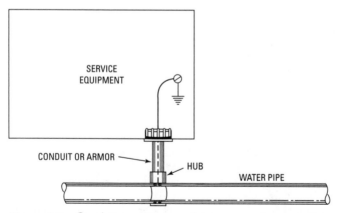

SERVICE
EQUIPMENT

CONDUIT OR ARMOR

HUB

WATER PIPE

Figure 6.8 Conduit or armor used to protect the grounding wire shall be bonded to the grounding electrode and to the service-entrance enclosure.

TWO-WIRE NM CABLE
WITH GROUND

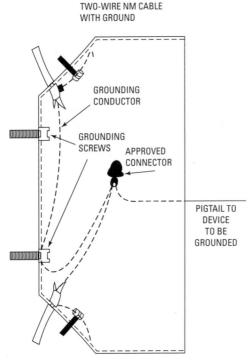

GROUNDING
CONDUCTOR

GROUNDING
SCREWS   APPROVED
CONNECTOR

PIGTAIL TO
DEVICE
TO BE
GROUNDED

Figure 6.9   The proper method of grounding boxes and carrying the grounding conductor on to the device to be grounded.

All paint or foreign substances must be removed from the area of grounding connections. This requirement is nothing more than common sense.

Isolated Grounding

Isolated grounding systems (which are used for many computer installations) are simply parts of the building's grounding system that are separated from the normal grounding system. *Every* item that is required to be grounded or bonded under the general grounding rules must be bonded or grounded even when it is connected to the isolated grounding system.

All pipes, boxes, enclosures, etc. that contain isolated grounding systems must be bonded to the normal grounding system; only grounding terminals on receptacles are to be connected to isolated grounding systems.

The isolated grounding system can remain completely separate from the normal grounding system, *except* the main connection to the grounding electrode(s) must be made at the same spot at which the main grounding system connects to the grounding electrode(s).

Isolated grounding conductors (which are always insulated, to keep them separate from the normal grounding system) are allowed to pass through any box, enclosure, or cabinet in a building without being bonded to anything, as long as they are connected to the main grounding electrode as mentioned above.

The sizes of isolated grounding conductors must be the same size as that required for a normal grounding conductor.

Refer to *Sections 250.146(D)* and *406.2(D)* of the NEC to find these rules.

Other Requirements

All instruments, etc. in or on switchboards must be grounded.

The grounding conductor for instruments, etc. must be at least No. 12 copper or No. 10 aluminum.

Where the primary voltage of instrument transformers exceeds 300 volts, and the transformers are accessible to unqualified persons, they must be grounded. If unaccessible to unqualified persons, they need not be grounded unless the primary voltage is over 1000 volts.

Instrument transformer cases must be grounded if they are accessible to unqualified persons.

High-voltage systems, where required to be grounded, must meet the same requirements for grounding as low-voltage systems (see *Sections 490.36, 490.37,* and *490.74*).

Surge Arrestors (*Article 280*)

Surge arrestors must be connected to all ungrounded conductors.

The rating of a surge arrestor (for circuits over 1000 volts) must be at least 125% of the operating voltage of the circuit it protects.

Surge arrestors can be installed indoors or outdoors. Unless specifically approved for such installations, they must be located where they are not accessible to unqualified persons.

Conductors connecting the surge arrestor to the system being protected must be as short as possible.

All surge suppressors should be connected according to the manufacturer's directions.

All surge suppressors must have a connection to one of the following:

The equipment grounding terminal.

The grounded service conductor.

The grounding electrode conductor.

The grounding electrode.

For services of 1000 volts or less, the minimum size for wires connecting surge arrestors is No. 14 copper or No. 12 aluminum.

For services of over 1000 volts, the minimum size for wires connecting surge arrestors is No. 6 copper or No. 6 aluminum.

When circuits are supplied by 1000 volts or more, the grounding conductor from a surge arrestor that supplies a transformer feeding a secondary distribution system must have one of the following:

If the secondary has a connection to a metal underground water piping system, the surge arrestor must have a connection to the secondary neutral.

If the secondary does *not* have a connection to a metal underground water piping system, the surge arrestor must be connected to the secondary neutral through a spark-gap device. See *Section 280.24(B)* for specifications of the spark-gap device.

Other connections, besides those mentioned immediately above, can be made by special permission.

Chapter 7

Wiring Requirements

Wiring Methods (*Article 300*)

General

The requirements of this section of the NEC are for wiring systems 600 volts or less and don't apply to the internal conductors of motors, controllers, etc.

Single conductors are permitted only where they are part of a standard wiring method.

All ungrounded conductors, neutral conductors, and equipment-grounding conductors of one circuit must be run in the same raceway, cable, cable tray, trench, or cord.

AC or dc circuits of voltages not over 600 volts are allowed in the same raceway, cable, or enclosure. But all wires in the raceway, cable, or enclosure must have insulation sufficient for the highest voltage present in the raceway, cable, or enclosure. A 120-volt circuit can occupy the same raceway with a 600-volt circuit (ac or dc), as long as all conductors in the raceway are insulated for 600 volts.

Circuits over 600 volts and circuits 600 volts or less can't occupy the same raceway, cable, or enclosure. Exceptions are made for lighting fixture ballast wires and control instrument wires.

Protection of Conductors

All conductors must be protected from physical damage.

When cables or raceways are installed through wood structural members, the edge of the bored holes must be at least $1\frac{1}{4}$ inch from the edge of the wood framing member.

Where the clearance specified above is not possible, a $\frac{1}{16}$-inch steel plate must be installed to cover the area of the wiring (Figure 7.1). Note that in the "real world," this requirement is often "winked at." For example, standard 2 × 4s are $3\frac{1}{2}$ inches wide. To keep the $1\frac{1}{4}$-inch clearance requirement, a hole no larger than 1 inch must be drilled *exactly* in the center of the 2 × 4. This is often not the case, and very few inspectors would require the steel plate mentioned here, unless the bored hole was very sloppy.

Notches can be cut in wood, if the strength of the structure won't be jeopardized. Wiring can be laid in the notches, and a $\frac{1}{16}$-inch steel plate must be installed to cover the area.

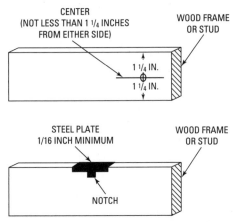

Figure 7.1 Proper drilling and notching procedures.

The steel plate mentioned above for covering notches is not required where the wiring method is one of the following:

Rigid metal conduit.

Intermediate metal conduit.

Electrical metallic tubing.

Rigid nonmetallic conduit.

Where non–metal-sheathed cables go through metal framing members (usually metal studs), a grommet or bushing must protect the hole that the cable passes through.

In locations where nails or screws are likely to damage non–metal-sheathed cables or electrical nonmetallic tubing, a 1/16-inch steel plate must be installed to cover the area.

The steel plate mentioned above is not required where the wiring method is one of the following:

Rigid metal conduit.

Intermediate metal conduit.

Electrical metallic tubing.

Rigid nonmetallic conduit.

Underground Wiring

Directly buried cables or conduits must meet the depth requirements given in *Table 300.5*.

Cables installed underground beneath a building must be in a raceway the full length of the building.

Conductors or cables directly buried must be protected from below where they emerge from the ground to a point 8 feet above the grade level. The protection can be provided by an enclosure or raceway.

Conductors that enter a building must be protected up to the point of entrance.

Directly buried conductors are allowed to be spliced without a junction box. However, the splicing method must be approved for the intended use.

Care must be taken so that backfill won't damage conductors. When necessary, a layer of sand or gravel should be placed over the conductors to protect them.

Conduits or raceways that could allow moisture to enter a building must be sealed.

Cables that terminate underground with directly buried wiring protruding from them must be terminated with a bushing or an equivalent sealing compound.

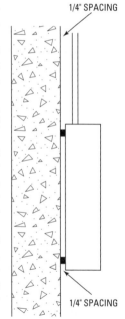

1/4" SPACING

1/4" SPACING

Figure 7.2 Air gap required in wet or frequently washed locations.

All conductors of the same circuit must be installed in the same raceway, or, for open conductors or cables, close to each other in the trench.

Parallel conductors are allowed, but each raceway must contain all the conductors of the same circuit.

All metal raceways, cables, and fittings must be suitable for the area in which they are installed. If corrosive conditions exist, raceways must be covered with a corro-sion-resistant coating or other suitable protection.

In wet locations, and in locations where the walls are frequently washed (dairies, laundries, etc.), all raceways, cables, enclosures, etc. must be installed with an air space of at least ¼ inch between them and the surface they are mounted on (Figure 7.2).

The air space requirement mentioned above also applies when raceways, cables, or enclosures are to be mounted on absorbent materials such as damp paper or wood.

Raceways

Air flow through a raceway must be prevented (usually by sealing) if the two ends of a run are exposed to noticeably different temperatures. This commonly applies to commercial freezer installations.

Raceways or cable trays that contain electric wiring can't also contain any other type of piping—steam, gas, air, drainage, etc.

All raceways, cables, and boxes must be electrically and mechanically joined together, except as allowed for nonmetallic boxes.

All raceways, cables, and boxes must be securely supported.

Raceways are not allowed to support other raceways, cables, or nonelectric equipment, unless identified as suitable for the use. Class II cables can also be supported, if they are used only for equipment control circuits.

Suspended ceiling support wires can be used to support equipment only if they are not part of a fire-rated floor or roof assembly. Furthermore, they can only be used to support branch-circuit wiring associated with equipment that is somehow attached to the suspended ceiling.

No splices are allowed in raceways.

The continuity of a grounded conductor can't be dependent on device connections. Neutrals can't "feed through" a wiring device, but must be made up with pigtail connections.

At least 6 inches of free conductor must be left at every box, fitting, or other splice point.

The above requirement does not apply to conductors that don't splice or terminate in the box, but rather feed through.

Boxes

Boxes or fittings are required at every splice or pull point in raceways. Wireways, gutters, and cable trays are excepted.

Boxes or fittings are required at every splice or pull point for AC, MC, MI, NM, or other cables. Certain exceptions exist. If in doubt, refer to *Section 300.15(A)* through (M), exceptions.

Fittings, etc. can be used only with the system they were designed for. For example, no greenfield fittings can be used on conduit.

A box or fitting with a separately bushed hole for each conductor must be used where a change is made between raceway or cable systems and open or knob-and-tube wiring methods (Figure 7.3).

Where a raceway terminates behind a switchboard, a bushing on the end of the raceway can be used instead of the box with bushed holes mentioned above.

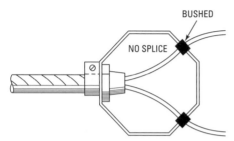

Figure 7.3 Method of transferring from cable to concealed knob-and-tube wiring.

Conductors

The number and size of conductors in raceways must not be too great to allow for heat dissipation and ease of installation and removal.

Raceways must be completely installed before conductors are pulled into the raceway system.

Conductors that are installed in vertical raceways must be supported. The support requirements are given in *Table 300.19(A)*.

Conductors of ac systems must be grouped together in circuits to avoid induction heating.

Where single ac conductors are used, the inductive effects must be avoided by use of the following methods:

1. Cutting notches in the metal between the holes the conductors pass through.

2. Running the conductors through a box or enclosure wall made of insulating material.

Openings around electrical penetrations in fire-rated walls, ceilings, floors, etc. must be properly sealed against the spread of fire.

No types of wiring are allowed in ducts that transport dust, loose stock, or flammable vapors.

The following wiring methods can be used in ducts or plenums used for environmental air:

Rigid metal conduit.

Intermediate metal conduit.

Electrical metallic tubing.

Flexible metallic tubing.

MI cable.

MC cable with a nonmetallic covering.

Flexible metal conduit, no longer than 4 feet.

Flexible liquid-tight metal conduit, no longer than 4 feet.

Equipment and devices are allowed in environmental air and plenum spaces only when necessary.

The following wiring methods can be used in spaces other than ducts or plenums used for environmental air:

Rigid metal conduit.

Intermediate metal conduit.

Electrical metallic tubing.

Flexible metallic tubing.

MI cable.

MC cable with a nonmetallic covering.

Flexible metal conduit.

Flexible liquid-tight metal conduit, no longer than 6 feet.

Type AC cable, or other listed cables.

Totally enclosed, nonventilated busway, with no provision for plug-ins.

Other types of conductors and cables can be installed in spaces other than ducts or plenums used for environmental air using the following methods:

Cable tray systems, solid bottom with solid metal covers, and only where accessible.

Surface metal raceways or metal raceways with removable covers, only where accessible.

Electrical equipment with metal enclosures or electrical equipment in listed nonmetallic enclosures can be installed in the places mentioned above. Integral fan systems can also be installed in such places.

Wiring under raised floors in data processing areas must meet the requirements for such areas. See *Article 645* and Chapter 32.

Temporary Wiring (*Article 527*)

Temporary wiring is allowed for the length of time of construction, demolition, remodeling, etc., or for a maximum of 90 days for Christmas lighting, carnivals, etc.

Temporary wiring may also be used for testing, experimental, and developmental work.

Temporary wiring must be removed as soon as the purpose for its installation is completed.

Temporary services must meet the same requirements as for permanent services.

Overhead temporary wiring may not be supported from vegetation, such as trees.

When the voltage to ground is not greater than 150 volts, feeders can be run as open conductors. In such cases, they must be supported on insulators every 10 feet or less.

When the voltage to ground is not greater than 150 volts, branch circuits can be run as open conductors. In such cases, they must be supported on insulators every 10 feet or less. They may *not* be laid on floors.

Full overcurrent protection must be provided for temporary wiring, the same as for permanent wiring.

Cords and cables can be used for temporary branch circuits, but the cord or cable must be listed for hard usage.

All receptacles must be of the grounding type.

Unless run in metal raceway or metal-covered cable, all branch circuits must have an equipment-grounding conductor.

Receptacles can't be connected to the same circuits that supply temporary lighting.

All lamps for general lighting must be equipped with a guard or be in a fixture.

Metal-cased lighting sockets can't be used unless the sockets are grounded.

Boxes are not required for splices in multiconductor cords or cables on construction sites.

A box, fitting, etc. with a bushed hole for each conductor must be used where a change is made to or from raceway systems to open wiring.

Flexible cords and cables must be protected from damage. Pinch points, such as passage through doorways, must be protected.

All single-phase, 125-volt, 15- and 20-ampere receptacles on construction sites must be ground-fault protected.

A testing system must be enforced at each construction site where temporary wiring is used. The following tests must be performed on all cord sets, receptacles, and cord-and-plug connected equipment:

> All equipment-grounding conductors must be checked for continuity.

Each receptacle and plug must be checked to verify that the equipment-grounding conductor is correctly attached.

The above-mentioned tests must be done:

Before the first use of temporary wiring at the construction site.

If there is evidence of damage.

After repairs, before return to service.

No more than once every three months.

The tests must be recorded, and the record of the tests must be available for the local inspector to review.

Fencing, or other suitable means of guarding, must be installed around temporary wiring operating at over 600 volts.

Conductors for General Wiring (*Article 310*)

All conductors must be insulated, unless specifically permitted to be bare.

Conductors (copper, aluminum, or copper-clad aluminum) sized 1/0 and larger are allowed to be installed in parallel (meaning electrically joined at each end), forming the equivalent of one single conductor. When this is done, it must be done for all conductors in the circuit, not one or two conductors in parallel and the others not in parallel.

Conductors in parallel must have the following characteristics:

They must be the same length.

They must be made of the same material, with the same insulation.

They must have the same circular mil area.

They must be terminated in the same way.

When they are run in separate raceways, the raceways must have the same characteristics.

Certain specialized systems can also have paralleled conductors (see *Section 310.4*, exceptions).

The minimum sizes of wires for specific voltages are listed in *Table 310.5*.

Solid dielectric conductors operating at over 2000 volts (for permanent installations only) must be shielded and have ozone-resistant insulation. All metallic shields must be grounded.

Direct-burial conductors must be identified for such use. Cables rated over 2000 volts must be shielded. The metal sheath or armor must be grounded.

Conductors used in wet locations must be listed for this purpose.

Cables used in wet locations must be listed for this purpose.

Cables used for direct burial must be listed for this purpose.

Conductors that are exposed to oils, vapors, gases, and other corrosive substances must be approved for this purpose.

Conductors can't be used where their temperature limit can be exceeded.

Identification of Conductors

Grounded circuit conductors (neutrals) No. 6 or smaller must be identified by being colored white or natural (light) gray.

An external ridge can be used to identify grounded conductors of multiconductor cables, No. 4 or larger.

Bare, covered, or insulated conductors are allowed as equipment-grounding conductors.

A conductor No. 6 or larger can be identified as a grounded conductor by a permanent white marking where exposed. This marking can be made with tape, paint, or both.

Ungrounded conductors must be identified by having colors besides white, green, or natural gray.

Generally, the ampacities of conductors are found in *Tables 310.16, 310.17, 310.18, 310.19, 310.20,* and *310.21* for common voltages, and *Tables 310.63* through *310.86* for higher voltages. Ampacities can also be calculated, under engineering supervision.

Section 310.15 is significant and should be reviewed.

Table 310.16 is the most commonly used, giving the ampacities of the most common types of conductors in cables or raceways.

Terminations for circuits that are rated 100 amps or less, and that use conductors from No. 14 through No. 1, are limited to 60° Celsius. Conductors that have higher temperature ratings (such as the commonly used THHN conductors) can be used for these circuits, but the ampacity of such conductors must be determined using the 60° columns of *Tables 310.16* through *310.19*.

If the termination devices for the circuits mentioned above are listed for operation at higher temperatures, the conductors can also have their ampacity calculated at the higher temperatures.

Terminations for circuits that are rated over 100 amps and that use conductors larger than No. 1 are limited to 75° Celsius. Conductors that have higher temperature ratings (THHN, XHHW, etc.) can be used for these circuits, but the ampacity of such conductors must be determined using the 75° columns of *Tables 310.16* through *310.19*.

If the termination devices for the circuits mentioned above are listed for operation at higher temperatures, the conductors can also have their ampacity calculated at the higher temperatures.

Wire connectors (such as wire nuts or lugs) must have temperature ratings equal to the temperature at which the conductor's ampacity was calculated. For example, in calculating the ampacity of a No. 8 conductor at 75° C, any splicing connector (such as a wire nut) used on those conductors must have a temperature rating of at least 75° C.

Chapter 8

Wiring in Cable

Cable Tray Systems (*Article 392*)

Uses and Locations

The following wiring methods can be used for installation in cable trays:

Mineral-insulated metal-sheathed cable.

MC cable.

Power-limited tray cable.

Nonmetallic-sheathed cable (shielded or unshielded).

Multiconductor service-entrance cable.

Multiconductor underground feeder and branch-circuit cable.

Power and control tray cable.

Fire alarm cables.

Optical fiber cables.

Other approved cables.

Any approved conduit or raceway, with the conductors enclosed therein.

The following cables are allowed to be installed in ladder, ventilated trough, and 3-, 4-, or 6-inch ventilated channel-type cable trays in industrial locations where only qualified persons will have access to the cable trays:

Single conductors 1/0 AWG and larger. Single-conductor cables 1/0 through 4/0 must be in a ventilated cable tray or a ladder-type cable tray with a maximum rung spacing of 9 inches. Sunlight-resistant cables must be used where exposed to the sun.

Multiconductor-type MV cables. Sunlight-resistant cables must be used where exposed to the sun.

Cable trays can be used as equipment-grounding conductors where only qualified persons can service the system.

Cable trays can be used in hazardous locations but must contain only cables approved for such areas.

In corrosive areas that require voltage isolation, nonmetallic cable trays can be used.

Cable trays can be used in environmental air spaces but must contain only cables approved for such areas.

Cable trays are *not* permitted in hoistways or where they may be subjected to physical damage.

Installation

Cables rated over 600 volts can't be placed in cable trays with cables rated 600 volts or under, except for type MC cables, or unless divided by barriers.

Cable trays must be installed as a complete system. Any field bends or modifications must not compromise the electrical and mechanical continuity of the system (Figure 8.1).

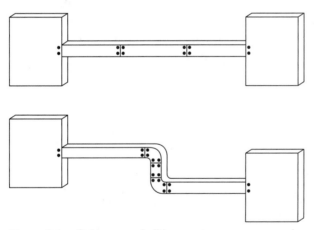

Figure 8.1   Cable trays shall be continuous as a complete system from the point of origin to the point of termination.

Complete runs of cable trays must be completed before the installation of any conductors.

Cables must be supported where they enter or leave a cable tray.

Multiconductor cables rated 600 volts or less can be installed in the same cable tray.

Covers or enclosures must be compatible with the cable tray.

Cable trays can penetrate walls, floors, partitions, etc., including fire-rated walls and floors, when the openings are sealed with a fire-rated compound (Figures 8.2, 8.3).

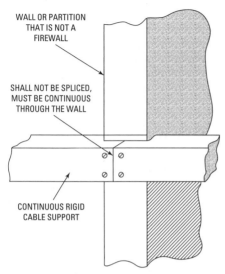

Figure 8.2 Cable trays shall be continuous through walls or partitions.

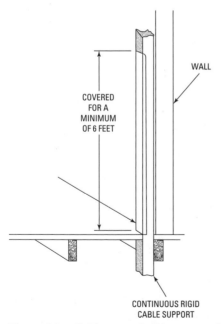

Figure 8.3 Cable trays shall be covered to a minimum height of 6 feet when run through floors.

Cable trays must be exposed and accessible. Sufficient space for maintenance must be provided.

Metal cable trays must be grounded.

Steel or aluminum cable trays can be used as equipment-grounding conductors. In these cases, the following conditions must be met:

The cable tray must be identified for grounding.

The cross-sectional area of the cable tray must comply with *Table 392.7(B)*.

All sections and fittings must be marked, showing the cross-sectional areas.

All sections and fittings must be bonded, using jumpers or mechanical connectors.

Cable Installation

Accessible splices are allowed in cable trays, but they must be made with approved devices.

Cables must be securely fastened to transverse members of the tray.

When bushings are installed on raceways feeding conductors into the cable tray, a box is not required.

When parallel conductors are installed, they must be bound together in circuit groups to prevent inductive effects.

Single conductors should also be grouped together (Figure 8.4).

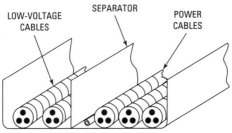

Figure 8.4   Separators shall be used where required.

Number of Cables

The number of multiconductor cables under 2000 volts allowed in a cable tray is indicated in *Table 318.9*, and the following restrictions apply:

For ladder or ventilated trays—if all cables are 4/0 or larger, the sum of all cable diameters can't exceed the tray width, and the cables must be installed in a single layer.

Multiconductor cables can fill up to 50% of the cross-sectional area of ladder or ventilated cable trays.

For solid-bottom trays—if all cables are 4/0 or larger, the sum of all cable diameters can't exceed 90% of the tray width, and the cables must be installed in a single layer.

Multiconductor cables can fill up to 40% of the cross-sectional area of solid-bottom cable trays.

Multiconductor cables can fill ventilated channel cable trays up to the following limits:

3-inch channels—1.3 square inches.

4-inch channels—2.5 square inches.

6-inch channels—3.8 square inches.

The amount of single-conductor cable under 2000 volts allowed in a cable tray is given in *Table 318.10*, and is subject to the the following restrictions:

For ladder or ventilated trays—if all cables are 1000 kcmil or larger, the sum of the cable diameters can't exceed the tray width.

For ladder or ventilated trays—if all cables are between 1/0 and 4/0, the sum of the cable diameters can't exceed the tray width, and the cables must be installed in a single layer.

Single-conductor cables in ventilated-channel cable trays (4-inch or 6-inch)—the sum of the cable diameters can't be greater than the inside width of the channel.

Ampacities of Conductors

The ampacities of conductors under 2000 volts in cable trays are based on the following:

Multiconductor cables—*Tables 310.16* and *310.18*, and *Section 310.15*. If the cable tray is covered for more than a 6-foot length, the ampacities can be only 95% of the values given in *Tables 310.16* and *310.18*. If the cables are installed in a single layer in open trays, with at least one cable diameter between cables, the ambient correction factors shown at the bottom of *Table 310.16* can be used.

Single-conductor cables (including conductors that are triplexed or quadruplexed):

If 600 kcmil or larger—75% of the values in *Tables 310.17* and *310.19*.

If between 1/0 and 500 kcmil—65% of the values in *Tables 310.17* and *310.19*, and Note 8 of the tables. If the cable tray is covered for more than a 6-foot length, the ampacities can be only 60% of the values in *Tables 310.16* and *310.18*.

If 1/0 and larger, laid in a single layer in open trays, and with at least one cable diameter between cables—the ampacities of *Tables 310.17* and *310.19* apply.

Where single conductors are installed in a triangular group in open trays, with a spacing of 2.15 conductor widths between circuits, the ampacity of 1/0 and larger cables is the same as that for messenger-supported conductors.

The ampacities of conductors over 2000 volts in cable trays are based on the following:

Multiconductor cables—*Tables 310.75* and *310.76*. If the cable tray is covered for more than a 6-foot length, the ampacities can be only 95% of those listed in *Tables 310.75* and *310.76*. If the cables are installed in a single layer in open trays, with at least one cable diameter between cables, the ampacities of *Tables 310.71* and *310.72* should be used.

Single-conductor cables (including conductors that are triplexed or quadruplexed):

If 1/0 or larger—75% of the values in *Tables 310.69* and *310.70*. If the cable tray is covered for more than a 6-foot length, the ampacities can be only 70% of those listed in *Tables 310.69* and *310.70*.

If the cables are spaced at least one cable diameter apart, the ampacities of cables 1/0 and larger are determined according to *Tables 310.69* and *310.70*.

Where single conductors are installed in a triangular group in open trays, with a spacing of 2.15 conductor widths between circuits, the ampacity of 1/0 and larger cables can be up to 105% of the value in *Tables 310.71* and *310.72*.

Open Wiring on Insulators (*Article 398*)

Uses Permitted

Open wiring is allowed for circuits 600 volts or less for industrial or agricultural establishments, indoors or outdoors, for services, and in places where it could be subjected to corrosive vapors.

Installation

Any of the conductor types of *Article 310* can be used.

The ampacities of open conductors can be found in *Tables 310.17* and *310.19*.

The conductors must be rigidly supported as follows:

Within a tap or splice—6 inches (Figure 8.5).

Within a dead-end connection (receptacle or lampholder)— 12 inches.

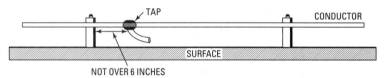

TAP

CONDUCTOR

SURFACE

NOT OVER 6 INCHES

Figure 8.5 Distance between a support and a tap should never exceed 6 inches.

Otherwise—4½ feet, or closer when necessary (Figure 8.6).

When noncombustible, nonabsorbent spacers are used every 4½ feet or less, No. 8 conductors can be supported every 15 feet. A spacing of 2½ inches between conductors must be maintained.

In mill buildings, if a 6-inch spacing is maintained between conductors, No. 8 and larger conductors can be run across open spaces without support.

In industrial buildings, conductors 250 kcmil or larger can be supported up to 30 feet across open spaces.

Supports must be mounted with at least ten-penny nails, or with screws that penetrate the wood surface to at least one-half the height of the knob.

Wires No. 8 or larger must be tie-wired to their supports with a wire having an insulation equal to that of the conductor.

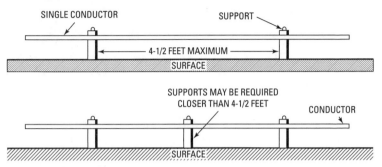

Figure 8.6 Spacing of supporting knobs.

Conductors can be enclosed in flexible nonmetallic tubing in dry locations that are not subject to damage. The tubing must be in pieces at least 15 feet long and must be supported every 4½ feet.

When conductors go through floors or walls, they must be run through an insulating, noncombustible tube (usually made of porcelain). If the tube is too short to reach all the way through the hole, an insulating sleeve can also be used. Each conductor must have its own tube.

Open conductors must be kept at least 2 inches from all other conductors or piping. If closer, they must be separated by firmly mounted insulating material. If tubes are used, they must be supported on each end.

As a suggestion, conductors should pass over, rather than under, other piping systems to prevent dripping, etc.

Drip loops must be made in conductors passing from wet or damp locations into dry locations. The loops should be made on the wet side. When entering, they should go slightly upward, so that water will drip into the wet area rather than the dry area.

Conductors within 7 feet of a floor must be protected by one of the following means:

By 1-inch-thick guard strips, placed next to the wiring.

By a running board at least ½ inch thick. The sides must be at least 2 inches high.

By the above method, with a cover that comes no closer than 1 inch from the conductor.

By being enclosed in:

Rigid metal conduit.

Intermediate metal conduit.

Rigid nonmetallic conduit.

Electrical metallic tubing.

When enclosed in raceways, care should be taken to group the conductors so that the current is approximately equal in both directions.

Conductors in attics or roof spaces that are accessible via a stairway or permanent ladder must be installed on floor joists, studs, or rafters up to 7 feet from the floor and protected by running boards that extend at least 1 inch on each side of the conductor.

Conductors in attics or roof spaces that are not accessible via a stairway or permanent ladder must be installed on floor joists, studs, or rafters. If there is less than 3 feet of headroom in the area, and the wiring is installed before the building is completed, it need not run along the floor joists, studs, or rafters.

If surface-mounted snap switches are mounted on insulators, no boxes are required.

Messenger-Supported Wiring (*Article 396*)

Uses and Locations

The following types of cables can be supported by messenger wires:

Mineral-insulated, metal-sheathed cables.

Metal-clad cables.

Multiconductor underground feeder and branch-circuit cables.

Power and control tray cables.

Other multiconductor, factory-assembled cables.

AC cables in dry locations.

In industrial establishments only—any of the conductor types in *Table 310.13* or *310.62*.

In industrial establishments only—MV cable.

Messenger-supported wiring is allowed in hazardous areas only when specifically permitted. See *Sections 501.4, 502.4, 503.3*, and *504.20*.

Messenger-supported wiring is *not* allowed in hoistways or where it is subject to physical damage.

Installation

The messenger wire must be supported at the ends and at locations in between to avoid tension on the wiring.

The conductors must be kept from touching the messenger supports, structural members, walls, or pipes.

Messenger wires must be grounded the same as equipment enclosures.

Any approved and insulated types of splices can be used.

Concealed Knob-and-Tube Wiring (*Article 394*)
Uses and Locations
Concealed knob-and-tube wiring may be used:

> In the hollow spaces of walls and ceilings.
>
> In unfinished attic and wall spaces (Figure 8.7).

Concealed knob-and-tube wiring may *not* be used:

> In commercial garages.
>
> In theaters and similar locations.
>
> In motion picture studios.
>
> In hazardous locations.
>
> In wall or ceiling spaces where loose, foam, or foamed-in insulation is present.

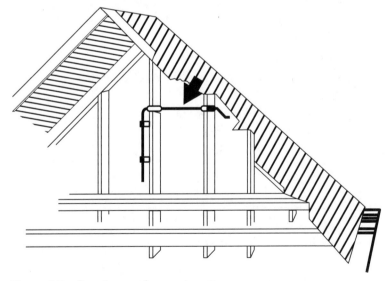

Figure 8.7 Running conductors in attics.

Installation

Any of the conductor types of *Article 310* can be used.

The conductors must be rigidly supported as follows:

Within a tap or splice—6 inches.

Otherwise—4½ feet.

When the conductors must be fished into finished walls, they can be enclosed in unbroken lengths of flexible nonmetallic tubing and don't require support.

Conductors must be tie-wired to the knobs with wires that have insulation equivalent to that of the conductor.

The conductors must be kept at least 3 inches from each other and 1 inch from building surfaces.

In areas (such as at panels) where space is limited, the conductors can be closer to each other than specified above, but must be enclosed in flexible nonmetallic tubing from the area that requires support back to the last support.

When conductors pass through floors, walls, or wood cross members, they must be protected by an insulating tube (made of porcelain). The tube must extend through the hole, and at least 3 inches beyond.

Open conductors must be kept at least 2 inches from all other conductors or piping. If closer, they must be separated by a firmly mounted insulating material. If tubes are used, they must be supported on each end.

Conductors in attics or roof spaces that are not accessible via a stairway or permanent ladder must be installed on floor joists, studs, or rafters up to 7 feet from the floor, and protected by running boards that extend at least 1 inch on each side of the conductor.

Conductors in attics or roof spaces that are not accessible via a stairway or permanent ladder must be installed on floor joists, studs, or rafters. If there is less than 3 feet of headroom in the area, and the wiring is installed before the building is completed, it need not run along the floor joists, studs, or rafters.

Approved types of splices must be used. In-line and strain splices are not allowed.

Integrated Gas Spacer Cable (*Article 326*)

Uses and Locations

Integrated gas spacer (IGS) cable may be used:

Underground, including direct burial.

For services.

As feeders.

As branch circuits.

Integrated gas spacer cable may *not* be used:

For interior wiring.

In contact with buildings.

Installation

The radius of any bend must be according to *Table 325.11*.

Individual runs of integrated gas spacer cable can't have more than four quarter bends (total of 360°).

A valve and cap (for checking and adding gas) must be provided for each length of cable.

Fittings must be designed for the cable, and must maintain the gas pressure.

The ampacity of IGS cable is according to *Table 325.14*.

Medium-Voltage Cable (*Article 328*)

Uses and Locations

Medium-voltage cable may be used:

On power systems up to 35,000 volts.

In wet or dry locations.

In cable trays, only as specifically permitted.

For direct burial.

As messenger-supported wiring.

Medium-voltage cable may *not* be used where exposed to direct sunlight.

The ampacity of MV cables must be according to *Tables 310.69* through *310.84*.

Flat Conductor Cable (*Article 324*)

Uses and Locations

Flat conductor cable may be used:

For branch circuits not over 300 volts between conductors or 150 volts to ground.

On finished floors.

On walls, when in surface metal raceways.

In damp locations.

On heated floors, when approved for this purpose.

Flat conductor cable may *not* be used:

Outdoors or in wet locations.

Where subject to corrosive vapors.

In hazardous locations.

In residential, school, and hospital buildings.

Installation

General-purpose branch circuits can't exceed 20 amperes; individual branch circuits can't exceed 30 amperes.

Floor-mounted FCC cable and its fittings must be covered with carpet squares no larger than 36 inches square. The squares must be secured with a release type of adhesive.

All connectors must be approved types, identified for their use. All bare ends must be insulated.

All floor-mounted FCC cables must have both top and bottom shields, covering and underneath their entire area.

All metal shields and associated fittings must be electrically continuous to the equipment-grounding conductor of the branch circuit.

All receptacles, housings, and devices must be identified for their use and connected to both the cable and the shields.

Connection to other wiring methods must be made via approved transition fittings.

FCC cable must be secured to floors by adhesive or mechanical means.

The crossing of only two cables is permitted.

FCC cables can pass over or under a flat communications cable.

Any part of an FCC cable installation that is higher from the floor than 0.09 inches must be tapered.

Cable runs can be added to, provided new cable connectors are used.

Proper polarization of the system must be maintained.

Mineral-Insulated Metal-Sheathed Cable (*Article 332*)

Uses and Locations

Mineral-insulated (MI) metal-sheathed cable may be used:

For service, feeders, or branch circuits.

For power, lighting, signal, or control circuits.

In dry, wet, or continuously moist locations.

Indoors or outdoors.

In exposed or concealed locations.

Embedded in masonry.

In any hazardous location.

In locations exposed to oil or gasoline.

Where exposed to corrosive conditions that don't harm its sheath.

Underground, where suitably protected (Figure 8.8).

Mineral-insulated metal-sheathed cable may *not* be used in destructively corrosive areas, unless protected.

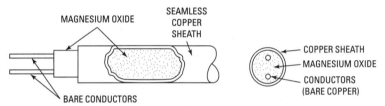

Figure 8.8   Construction of MI cable.

Installation

In wet locations and in locations where the walls are frequently washed (dairies, laundries, etc.), mineral-insulated metal-sheathed cables must be installed with an air space of at least ¼ inch between them and the surface they are mounted upon.

When MI cables are installed through wood structural members, the edge of the bored holes must be at least 1¼ inch from the edge of the wood framing member.

Where the clearance specified above is not possible, a ¹⁄₁₆-inch steel plate must be installed to cover the area of the wiring.

MI cables must be supported at least every 6 feet. This requirement is excepted when the cables must be fished in.

Bends must be made in such a way that the cable is not damaged. The minimum bend radius can't be less than five times the cable diameter (Figure 8.9).

Fittings must be approved for the purpose (Figure 8.10).

Where single conductors are used, inductive effects must be avoided by grouping all cables of circuits together, including neutral conductors of these circuits. Where these conductors enter metal enclosures, induction heating must be prevented by the following methods:

I. Cutting notches in metal boxes between the holes the conductors pass through.

2. Running the conductors through a box or enclosure wall made of insulating material.

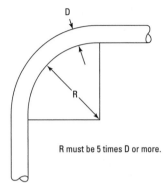

R must be 5 times D or more.

Figure 8.9   Radius of bends.

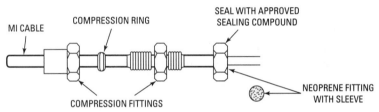

Figure 8.10   Fittings for MI cable.

At terminations, seals must immediately be made to prevent moisture from entering the cable, and the conductors must be insulated (Figure 8.11).

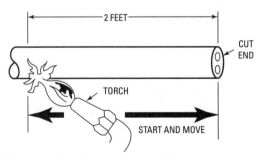

Figure 8.11   One method of driving out moisture.

Armored Cable (*Article 320*)
Uses and Locations
Armored cable may be used:

> For branch circuits and feeders.
>
> In exposed or concealed locations.
>
> In identified cable trays.
>
> In dry locations.
>
> Embedded in masonry.
>
> Fished in the voids of block walls.
>
> For underplaster extensions.
>
> Underground in raceways.

Armored cable may *not* be used:

> In theaters and similar locations.
>
> In motion picture studios.
>
> In hazardous locations, except as permitted (see *Sections 501.4(B), 502.4(B), and 504.20*).
>
> In areas where corrosive vapors exist.
>
> On cranes or hoists.
>
> In storage battery locations.
>
> In hoistways or elevators.
>
> In commercial garages.

Installation
AC cables must be supported within 12 inches of every box, cabinet, or fitting; and every 4½ feet thereafter.

The above requirement is excepted in the following cases:

1. Where the cable is fished into place.
2. Lengths of 2 feet are allowed at terminals where flexibility is required.
3. In lengths up to 6 feet for connection to light fixtures (or other items) in accessible ceilings.

Bends must not be made so tightly that the cable can be damaged.

Approved insulating bushings must be used at all connections, unless the box or fitting provides equal protection.

Approved connectors must be used for AC cables.

When AC cables are installed through wood structural members, the edge of the bored holes must be at least 1¼ inch from the edge of the wood framing member (Figure 8.12).

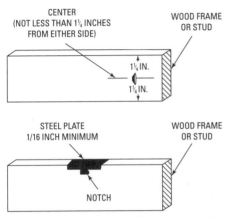

Figure 8.12.   Proper drilling and notching procedures.

Where the clearance specified above is not possible, a ¹/₁₆-inch steel plate must be installed to cover the area of the wiring.

Exposed runs of AC cables must follow the surface of the building or running boards, except as follows:

1. Lengths up to 2 feet where flexibility is required.

2. On the underside of floor joists in basements. Cables must be supported at every joist and not be subject to damage.

3. In lengths up to 6 feet for connection to light fixtures (or other items) in accessible ceilings.

AC cables in attics or roof spaces that are accessible via a stairway or permanent ladder and installed on the top of floor joists or on the face of studs or rafters up to 7 feet from the floor must be protected by running boards that extend at least 1 inch on each side of the conductor. When attics are not accessible via a stairway or permanent ladder, cables must be protected only within 6 feet of the opening to the attic.

When run on the sides of floor joists, studs, or rafters, no running boards are required.

Metal-Clad Cable (*Article 330*)
Uses and Locations
Metal-clad (MC) cables may be used:

> For branch circuits and feeders.
>
> For power, lighting, control, or signal circuits.
>
> For direct burial, where identified for such use.
>
> In exposed or concealed locations.
>
> In cable trays.
>
> As open runs of cable.
>
> As aerial cable on a messenger.
>
> As specifically permitted for hazardous locations.
>
> In dry locations.
>
> In raceways.
>
> In wet locations in any of the following cases:
>
>> The metal covering is impervious to moisture.
>>
>> A lead sheath is provided.
>>
>> The conductors are approved for this use.

Metal-clad cables may *not* be used where exposed to destructive corrosive vapors.

If single-conductor cables are used, the phase conductors must be grouped together, along with a neutral if used. This is done to minimize voltages induced into the sheath.

Installation
Metal-clad cables must be supported at least every 6 feet.

Metal-clad cables in cable trays must comply with *Article 392*.

Bends must be made in such a way that cables will not be damaged.

The minimum radius for various types of cables is as follows:

> Cables not more than ¾ inch in diameter—ten times the cable diameter.
>
> Cables between ¾ and 1½ inches in diameter—twelve times the cable diameter.
>
> Cables greater than 1½ inches in diameter—fifteen times the conductor diameter.

Cables with interlocked armor or corrugated sheath—seven times the sheath's external diameter.

Cables with shielded conductors—twelve times the shield diameter, or seven times cable diameter, whichever is greater.

Fittings must be identified for their use.

Where single conductors are used, inductive effects must be avoided by the following methods:

Cutting notches in metal boxes between the holes the conductors pass through.

Running the conductors through a box or enclosure wall made of insulating material.

The ampacity of MC cables must be according to *Tables 310.16* through *310.19.*

Nonmetallic-Sheathed Cable (*Article 334*)
Uses and Locations
Nonmetallic-sheathed (NM) cable can be used in:

One- and two-family dwellings, multifamily dwellings, and other structures.

Identified cable trays.

Exposed or concealed locations.

Dry locations.

Air voids of masonry block walls.

Additionally, type NMC (corrosion-resistant) can also be used in:

Damp or wet locations.

IMasonry block walls.

Shallow masonry chase, protected with a $\frac{1}{16}$-inch steel plate, and covered with plaster, etc.

NMC cable *can't* be used:

In any structure that exceeds three stories above grade. The first level, or story, of a building is a level that has half or more of its exterior wall surface at or above finished grade. If the first floor of a building is used only for parking or storage, it does not have to be counted as one of the three permitted floors.

As service-entrance cable.

In commercial garages.

In theaters and similar locations, except where specifically permitted.

In motion picture studios.

In storage battery areas.

In hoistways.

Embedded in masonry.

In hazardous locations, except as specifically permitted.

Type NMC cable *can't* be installed in the following areas:

Where exposed to corrosive vapors.

Embedded in masonry.

In a shallow chase in masonry.

Installation

NM cables must be supported within 12 inches of every box, cabinet, or fitting; and every 4½ feet thereafter.

The above requirement is excepted where the cable is fished into place.

Devices designed for the termination of NM cables without boxes are allowed if the cables are secured within 12 inches of the device. A 6-inch loop of cable must also be left inside the wall.

The cable must follow the surface of the building or running boards.

When necessary, the cable must be protected from damage by metal pipe or tubing. When passing through a floor, the cable must be protected at least 6 inches above the floor.

When NM cables are installed through wood structural members, the edge of the bored holes must be at least 1¼ inch from the edge of the wood framing member. Where the clearance specified above is not possible, a ¹⁄₁₆-inch steel plate must be installed to cover the area of the wiring.

Two conductor cables can't be stapled on edge. Passing through a joist, rafter, etc., is considered to be a support. Insulated staples must be used for cables with conductors smaller than No. 8. Cable ties are acceptable as a supporting means.

Cables with two No. 6 or three No. 8 conductors can be run across joists in unfinished basements without running boards but they must be secured at every joist.

Bends must not damage the cable. The minimum bending radius is five times the cable diameter.

In attics or roof spaces accessible via a stairway or permanent ladder, NM cables installed on the top of floor joists, or on the face of studs, or on rafters up to 7 feet from the floor must be protected by running boards that extend at least 1 inch on each side of the conductor. When attics are not accessible via a stairway or permanent ladder, cables must be protected only within 6 feet of the opening to the attic.

Switch, outlet, and tap boxes that are made of insulating material can be used without boxes for exposed wiring or rewiring work where the cables are concealed and fished into place. In these cases, the devices must fit closely around the cables and enclose all stripped parts of the cables. All conductors must terminate; none should be left unattached.

NM cables are allowed with nonmetallic boxes.

NM cables are allowed with devices with integral enclosures.

Type NMS

Type NMS is a nonmetallic sheathed cable that also has shielded communications conductors as part of the cable.

Uses and Locations

NMS cable may be installed:

> Where the operating temperatures don't exceed the rating of the cable.
>
> In cable trays.
>
> In raceways.
>
> Where permitted in hazardous locations.
>
> In agricultural buildings, as permitted.

NMS cable may *not* be installed:

> In any area not mentioned above.

Installation

Bends must not be so tight as to damage the cable. Five times diameter is the minimum radius.

Handling of the cable must not damage its covering.

Fittings must be identified for their use.

The wire shield must be bonded to the equipment served and at the power supply point. Bonding can be accomplished by fittings or by other means.

Service-Entrance Cable (*Article 338*)

Uses Permitted

For service entrances, branch circuits, or feeders.

Installation

When used as interior wiring, all of the cable conductors must have rubber or thermoplastic insulation.

Service-entrance (SE) cable with uninsulated neutrals (Figure 8.13) can't be used as interior wiring, unless it has an outer nonmetallic covering and operates at no more than 150 volts to ground. In these cases, it can be used for the following:

As a branch circuit that supplies a range, oven, cooktop, or dryer.

As a feeder to supply other buildings on the same premises.

SE cable can't be subjected to temperatures higher than the rating of the cable.

SE cable must be supported within 12 inches of every box, cabinet, or fitting, and every 4½ feet thereafter.

The above requirement is excepted where the cable is fished into place.

The cable must follow the surface of the building or running boards.

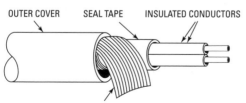

OUTER COVER   SEAL TAPE   INSULATED CONDUCTORS

BARE CONDUCTOR WRAPPED AROUND SEAL TAPE

Figure 8.13 SE cable with bare neutral.

When necessary, the cable must be protected from damage by metal pipe or tubing. When passing through a floor, the cable must be protected at least 6 inches above the floor.

When SE cables are installed through wood structural members, the edge of the bored holes must be at least 1¼ inch from the edge of the wood framing member. Where the clearance specified above is not possible, a ¹⁄₁₆-inch steel plate must be installed to cover the area of the wiring.

Cables with two No. 6 or three No. 8 conductors can be run across joists in unfinished basements without running boards, but they must be secured at every joist.

Bends must not damage the cable. The minimum bending radius is five times the cable diameter.

In attics or roof spaces accessible via a stairway or permanent ladder, SE cables installed on the top of floor joists, or on the face of studs, or on rafters up to 7 feet from the floor must be protected by running boards that extend at least 1 inch on each side of the conductor. When attics are not accessible via a stairway or permanent ladder, cables must be protected only within 6 feet of the opening to the attic.

Underground Feeder and Branch-Circuit Cable (*Article 340*)

Uses and Locations

Underground feeder (UF) and branch-circuit cable may be installed:

Underground, including direct burial.

As feeder or branch-circuit cable.

In wet or dry locations.

In corrosive locations.

Underground feeder and branch-circuit cable may *not* be installed:

In theaters.

In commercial garages.

As service-entrance cables.

In motion picture studios.

In storage battery areas.

In hoistways.

In hazardous locations.

Embedded in masonry.

Where directly exposed to sunlight.

Installation

If installed as single-conductor cables, the conductors of circuits must be placed close to each other in a trench.

Only multiconductor UF cables can be used in cable trays.

UF cable can't be subjected to temperatures higher than the rating of the cable.

UF cables must be supported within 12 inches of every box, cabinet, or fitting, and every 4½ feet thereafter.

The above requirement is excepted where the cable is fished into place.

The cable must follow the surface of the building or running boards.

When necessary, the cable must be protected from damage by metal pipe or tubing. When passing through a floor, the cable must be protected at least 6 inches above the floor.

When UF cables are installed through wood structural members, the edge of the bored holes must be at least 1¼ inch from the edge of the wood framing member. Where the clearance specified above is not possible, a ¹⁄₁₆-inch steel plate must be installed to cover the area of the wiring.

Cables with two No. 6 or three No. 8 conductors can be run across joists in unfinished basements without running boards, but they must be secured at every joist.

Bends must not damage the cable. The minimum bending radius is five times the cable diameter.

In attics or roof spaces accessible via a stairway or permanent ladder, UF cables installed on the top of floor joists, or on the face of studs, or on rafters up to 7 feet from the floor must be protected by running boards that extend at least 1 inch on each side of the conductor. When attics are not accessible via a stairway or permanent ladder, cables must be protected only within 6 feet of the opening to the attic.

Power and Control Tray Cable (*Article 336*)

Uses and Locations

Power and control tray cable may be installed:

In cable trays.

For power, lighting, or control circuits.

In raceways.

Supported by a messenger wire.

As permitted, in cable trays in hazardous locations.

For class I circuits.

Power and control tray cable may *not* be installed:

Where subject to damage.

As open cable on brackets or cleats.

Directly exposed to the sun.

For direct burial, unless specifically identified as suitable.

Installation
Bends must be made in such a way that the insulation is not damaged.

Nonmetallic Extensions (*Article 382*)
Uses and Locations
Nonmetallic extensions may be installed:

From an existing outlet on a 15- or 20-ampere circuit.

In exposed locations.

In dry locations.

In occupied buildings not exceeding three floors above grade.

For aerial cable—in industrial areas that require an extremely flexible wiring system (Figure 8.14).

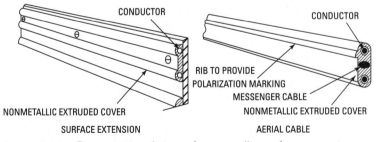

Figure 8.14 Cross-sectional view of nonmetallic surface extension cable and nonmetallic aerial cable.

Nonmetallic extensions may *not* be installed:

As an aerial cable substituting for other aerial wiring methods.

In unfinished attics, basements, or roof spaces.

Where voltage between conductors is over 150 volts for interior use, and 300 volts for aerial cable.

Where subject to corrosive vapors.

Run through a floor or partition.

Installation

Splices and taps are allowed only in approved fittings.

Each run must terminate in identified fittings.

Extensions can't be run within 2 inches of a floor.

Extensions must be secured every 8 inches. The first attachment can be 12 inches if an attachment plug is used to connect to the outlet (Figure 8.15).

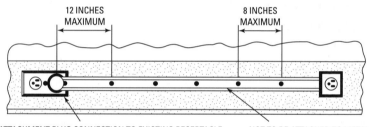

Figure 8.15   Supporting distances for nonmetallic surface extensions.

No run, no matter how short, can be without support.

Bends that alter the conductor spacing must be covered.

Aerial cables must be supported by messenger at each end and have intermediate supports of 20 feet or less.

Aerial cables must have at least 10 feet of clearance above pedestrian areas and 14 feet of clearance above areas subject to vehicular traffic.

Aerial cables must have 8 feet of clearance above working surfaces.

Aerial cables can support lighting fixtures when the load is not too much for the supporting messenger cable.

If the messenger cable is properly installed, it can serve as an equipment grounding conductor.

Flat Cable Assemblies (*Article 322*)
Uses and Locations
Flat cable (FC) assemblies may be installed:

> As branch circuits supplying lighting, small appliance, or small power loads.
>
> In exposed conditions.

Flat cable assemblies may *not* be used:

> Where exposed to damage.
>
> Where subject to corrosive vapors.
>
> In hazardous locations.
>
> In hoistways.
>
> In damp or wet locations.

Installation
Flat cable assemblies may be installed only in identified surface metal raceways (*Unistrut* or *Kindorf*).

> Splices can be made only in approved junction boxes.
>
> Taps must be made with identified devices.
>
> All dead ends must terminate in an identified end cap.
>
> Only identified fixture hangers can be used.
>
> Extensions must be made only from the ends of the runs, by approved means.
>
> Only identified fittings can be used.
>
> Support shall be as required for the specific raceways.
>
> Branch circuits can't be rated more than 30 amperes.
>
> When installed less than 8 feet from the floor, FC cables must have a protective cover.

Chapter 9

Wiring in Conduit

Electrical Metallic Tubing (EMT, Thin-Wall) (*Article 358*)

Uses and Locations

EMT may be installed in all locations *except:*

> Where it will be vulnerable to severe damage.
>
> In concrete or cinder fill. EMT may be installed in these locations if enclosed on all sides by 2 inches or more of concrete, or if the tubing is 18 inches or more below the fill (Figures 9.1, 9.2).

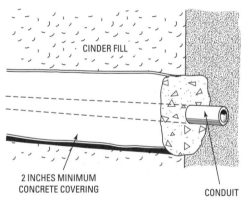

CINDER FILL

2 INCHES MINIMUM
CONCRETE COVERING

CONDUIT

Figure 9.1 Concrete covering for conduit under cinder fill.

> In any hazardous location, unless specifically permitted. *(Sections 502.4, 503.3, 504.20)*
>
> In concrete, earth, or corrosive areas, except when protected to the satisfaction of the local authorities.

Installation

EMT may not be threaded.

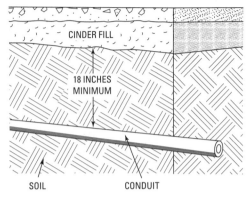

Figure 9.2 No concrete is required if conduit is buried a minimum of 18 inches below cinder fill.

Rain-tight (compression) fittings must be used in wet (outdoor) locations.

Bends must not crimp or flatten the tubing. Use only benders designed for the purpose.

No more than 360° of bend (equivalent to four quarter bends) is allowed between pulling points (Figure 9.3).

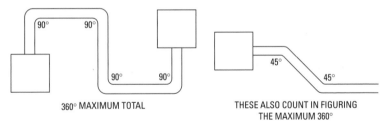

Figure 9.3 A maximum of 360° allowed between outlets and/or fittings.

All ends that are cut must be reamed.

EMT must be strapped (supported) within 3 feet of every box, cabinet, or fitting (condulet type), and every 10 feet thereafter.

If support is not possible within 3 feet of a box, cabinet, or fitting, an unbroken piece of EMT ("unbroken" meaning with no couplings in the section) may run 5 feet before it is supported.

Wire Fills

The sizes and numbers of conductors allowed in raceways are shown in the tables of Chapter 9, found at the back of the 2002 National Electrical Code.

Rigid Metal Conduit (RMC or Heavy-Wall) (*Article 344*)

Uses and Locations

Rigid metal conduit is acceptable for all uses in all locations, *with the following restrictions:*

When installed in contact with earth, concrete, or in corrosive areas, it must be suitably protected.

When installed in or under cinder fill, it must be enclosed on all sides by 2 inches or more of concrete, be 18 inches or more below the fill, or be suitably protected.

Installation

It is required that "where practical," contact between rigid metal conduit and different types of metals be avoided to eliminate or reduce galvanic action.

All ends that are cut must be reamed.

Bushings must be used to protect wires wherever rigid metal conduit enters a box, fitting, or enclosure. There are a few types of boxes, fittings, and enclosures that have built-in protection for wires. In these obvious cases, a bushing is not required.

Threadless fittings must be suitable for the area in which they are installed. Concrete-tight fittings must be used in concrete, and rain-tight (compression) fittings must be used in wet locations.

Running threads may not be used where the conduit connects to couplings.

Conduit may not be flattened when bent. Use benders designed for the purpose.

No more than 360° of bend is allowed between pulling points.

Conduit must be supported within 3 feet of every box, cabinet, or fitting (condulet type), and every 10 feet thereafter.

If only threaded couplings and proper supports are used, 1-inch conduit may be supported every 12 feet; 1¼-inch and 1½-inch conduit may be supported every 14 feet; 2-inch and 2½-inch conduit may be supported every 16 feet, and conduit 3-inch and larger may be supported every 20 feet.

Runs of conduit from industrial machinery that use only threaded couplings, and are supported both top and bottom, may be supported only every 20 feet in between.

Wire Fills
The sizes and numbers of conductors allowed in raceways are shown in the tables of Chapter 9, at the back of the 2002 NEC.

Intermediate Metal Conduit (IMC) (*Article 342*)

Uses and Locations
Intermediate metal conduit is acceptable for all uses in all locations, *with the following restrictions:*

> When installed in contact with earth, concrete, or in corrosive areas, it must be suitably protected.

> When installed in or under cinder fill, it must be enclosed on all sides by 2 inches or more of concrete, be 18 inches or more below the fill, or be suitably protected.

Installation
It is required, "where practical," that contact between intermediate metal conduit and different types of metals be avoided, to eliminate or reduce galvanic action.

All ends that are cut must be reamed.

Bushings must be used to protect wires wherever time intermediate metal conduit enters a box, fitting, or enclosure. There are a few types of boxes, fittings, and enclosures that have built-in protection for wires. In these obvious cases, a bushing is not required.

Threadless fittings must be suitable for the area in which they are installed. Concrete-tight fittings must be used in concrete, and rain-tight (compression) fittings must be used in wet locations.

Running threads may not be used where the conduit connects to couplings.

Conduit may not be flattened when bent. Use benders designed for the purpose.

No more than 360° of bend is allowed between pulling points.

Conduit must be supported within 3 feet of every box, cabinet, or fitting (condulet type), and every 10 feet thereafter.

If only threaded couplings are used and proper supports are used, 1-inch conduit may be supported every 12 feet; 1¼-inch and 1½-inch conduit may be supported every 14 feet; 2-inch and 2½-inch conduit may be supported every 16 feet, and conduit 3-inch and larger may be supported every 20 feet.

Runs of conduit from industrial machinery that use only threaded couplings, and are supported both top and bottom, may be supported only every 20 feet in between.

Wire Fills

The sizes and numbers of conductors allowed in raceways are shown in the tables of Chapter 9, at the back of the 2002 NEC.

Rigid Nonmetallic Conduit (RNC) *(Article 352)*

Uses and Locations

Rigid nonmetallic conduit is acceptable for all uses and locations, *except:*

> It may not be used in hazardous locations, except where specifically permitted (see *Sections 503.3(A), 504.20, 514.8, 505.8,* and *501.4(B),* exceptions).
>
> It may not be used to support lighting fixtures.
>
> It may not be used in locations where it will be subjected to physical damage.
>
> Rigid nonmetallic conduit may be used in theaters and similar locations only when encased in 2 inches or more of concrete.
>
> It may not be used in locations where the temperature is higher than the temperature for which the conduit is rated. The temperature rating should be stamped on the conduit every 10 feet.

Installation

All cut ends must be trimmed, removing rough edges.

Rigid nonmetallic conduit must be supported within 3 feet of every conduit termination. Thereafter, ½-inch, ¾-inch, and 1-inch rigid nonmetallic conduit must be supported every 3 feet; 1¼-inch, 1½-inch, and 2-inch conduit must be supported every 5 feet; 2½-inch and 3-inch conduit must be supported every 6 feet; 3½-inch, 4-inch, and 5-inch conduit must be supported every 7 feet; and 6-inch conduit must be supported every 8 feet.

Expansion joints must be installed in long runs of rigid nonmetallic conduit. Exactly how long a run requires an expansion joint is not specified in the code, but *Table 10* in NEC Chapter 9 shows the expansion characteristics of PVC conduit.

Bushings must be used in terminating rigid nonmetallic conduit, unless the box or termination fitting provides protection. (Virtually all do.)

Bends in rigid nonmetallic conduit must be made with benders designed for the purpose, and must not flatten or crimp the conduit.

No more than 360° of bend is allowed between pulling points.

Wire Fills

The sizes and numbers of conductors allowed in raceways are shown in the tables of Chapter 9, at the back of the 2002 NEC.

Flexible Metal Conduit (Greenfield or Flex) (*Article 348*)

Uses and Locations

Flexible metal conduit may *not* be used in the following locations:

In wet locations, unless lead-covered conductors are used, or unless approved for the specific installation.

In hoistways, except as specifically permitted (*Section 620.21(A)(1)*).

In storage battery rooms.

In hazardous locations, except as specifically permitted (*Sections 501.4(B)* and *504.20*).

Underground.

Embedded in concrete or aggregate.

Three-eighths-inch flexible metal conduit may be used only for underplaster extensions, enclosing motor leads, lighting fixture whips, on elevators (only where less than 6 feet long), and as factory-manufactured systems.

Installation

Flexible metal conduit must be supported within 12 inches of every box, cabinet, or fitting, and every 4½ feet thereafter. *Exception:* Flexible metal conduit does not have to be supported when used as fixture whips, fished into place, or in lengths less than 3 feet at terminals where flexibility is necessary.

Flexible metal conduit can be used for a grounding conductor only if the run is less than 6 feet long and uses fittings approved for grounding, and the circuit is protected at 20 amperes or less.

No more than 360° of bend is allowed between pulling points.
Angle connectors can't be used on concealed raceways.

Wire Fills

The sizes and numbers of conductors allowed in raceways are
shown in the tables of Chapter 9, at the back of the 2002 NEC.

Flexible Metallic Tubing (*Article 360*)

Uses and Locations

Flexible metallic tubing may be used for dry, accessible, protected
locations, and only for branch circuits less than 1000 volts, in
lengths of 6 feet or less.

Flexible metallic tubing may *not* be used:

In hoistways.

In storage battery locations.

In hazardous locations, except where specifically permitted.

Underground.

Embedded in concrete or aggregate.

In areas where it might be subjected to physical damage.

Installation

Three-eighths-inch flexible metallic tubing may be used only in environmental air ducts, plenums or spaces, or as lighting fixture whips.

Three-quarters-inch flexible metallic tubing is the largest size
permitted.

Flexible metallic tubing can be used for a grounding conductor
only if the run is less than 6 feet long and uses fittings approved for
grounding, and the circuit is protected at 20 amperes or less.

The minimum bend radii for fixed bends of flexible metallic tubing are 3½ inches for ⅜-inch flexible metallic tubing, 4 inches for
½-inch flexible metallic tubing, and 5 inches for ¾-inch flexible
metallic tubing.

The minimum bend radii for flexible bends are 10 inches for ⅜-
inch flexible metallic tubing, 12½ inches for ½-inch flexible
metallic tubing, and 17½ inches for ¾-inch flexible metallic
tubing.

Wire Fills

The sizes and numbers of conductors allowed in raceways are
shown in the tables of Chapter 9, at the back of the 2002 NEC.

Liquid-Tight Flexible Metallic Conduit (LFMC) (*Article 350*)

Uses and Locations

Liquid-tight flexible metallic conduit is permitted:

Where flexibility is required.

Where conditions in the area of installation require liquid-tight or vapor-tight protection, either exposed, concealed, or underground.

In hazardous locations (but *not* class I, division 1) and floating buildings where flexibility is necessary. See *Sections 501.4(B), 502.4, 503.3, 504.20, and 553.7(B).*

It is *not* permitted:

Where it may be subjected to physical damage.

Where the operating temperature might be higher than the rating of the liquid-tight conduit.

Installation

Three-eighths-inch liquid-tight flexible conduit may be used only in environmental air ducts, plenums or spaces, or as lighting fixture whips. LFMC of at least ½-inch must be used for all other installations.

Liquid-tight flexible metallic conduit must be supported within 12 inches of every box, cabinet, or fitting, and every 4½ feet thereafter. *Exception:* Liquid-tight flexible metallic conduit need not be supported when used as fixture whips, fished into place, or in lengths of less than 3 feet at terminals where flexibility is necessary.

Liquid-tight flexible metallic conduit can be used for a grounding conductor only if the run is less than 6 feet long and uses fittings that are approved for grounding, and the circuit is protected at 20 amperes or less for ⅜-inch and ½-inch LFMC, and at 60 amps or less for ¾-inch, 1-inch, and 1¼-inch LFMC.

No more than 360° of bend is allowed between pull points.

Angle connectors are not to be used for concealed installations.

Wire Fills

The sizes and numbers of conductors allowed in raceways are shown in the tables of Chapter 9, at the back of the 2002 NEC.

Liquid-Tight Flexible Nonmetallic Conduit (LFNC) (*Article 356*)

Uses and Locations

Liquid-tight flexible nonmetallic conduit may be used:

> Where flexibility is required.
>
> Where conditions in the area of installation require liquid-tight or vapor-tight protection, either exposed, concealed, or underground.
>
> Outdoors, when marked as suitable for that use.
>
> Directly buried, when marked as suitable for that use.

It may *not* be used:

> Where it may be subjected to physical damage.
>
> Where the operating temperature might be higher than the rating of the liquid-tight conduit.
>
> In hazardous locations.
>
> Where the voltage of the enclosed conductors exceeds 600 volts.

Installation

When liquid-tight nonmetallic conduit is used as a fixed raceway, it must be supported within 12 inches of every box, cabinet, or fitting. Thereafter, it must be supported every 4 feet.

It is permissible to fish liquid-tight nonmetallic conduit into place without support. Where flexibility is required, it may be supported up to 3 feet from a box or fitting, and it may be used unsupported in lengths up to 6 feet for fixture whips.

Runs of liquid-tight nonmetallic conduit can't have more than 360° of bends (equal to four quarter bends) between pull boxes, junction boxes, etc.

Three-eighths-inch liquid-tight flexible nonmetallic conduit may be used only for enclosing motor leads, as whips to fixtures or equipment (not over 6 feet long), or for electric-sign circuits (see *Section 600-31(a)*).

All fittings must be identified for use with liquid-tight flexible nonmetallic conduit.

Equipment-grounding conductors must be used.

Exterior equipment-grounding conductors may not be longer than 6 feet, and must be routed with the conduit.

All boxes and fittings must be bonded.

Wire Fills

The sizes and numbers of conductors allowed in raceways are shown in the tables of Chapter 9, at the back of the 2002 NEC.

Chapter 10

Raceways and Wireways

Surface Metal Raceways (*Article 386*)

Uses and Locations
May be used:

> In dry locations.
>
> As underplaster extensions, when identified as suitable for such use.
>
> Under raised floors (see *Section 645.5(D)(2)* of the NEC).

May *not* be used:

> Where the surface metal raceway might be subjected to severe physical damage.
>
> To enclose conductors that have a voltage of 300 V or higher between conductors, unless the metal is at least 0.040 inches thick.
>
> In areas where hazardous vapors are present.
>
> In concealed locations, except for underplaster extensions and under raised floors.
>
> In hazardous locations, except when enclosing conductors in Class I, Division 2 locations that don't carry enough power to ignite specific atmospheres (see *Section 504.4(B)*).

Installation
Only unbroken pieces of surface metal raceway may be extended through floors, walls, or partitions.

When divided raceways (usually one section for power wiring, and one or more sections for data or communications wiring) are used, the positions of the sections relative to one another (such as data in the left-hand section and power in the right-hand section) must be maintained throughout the installation, to avoid confusion.

Splices or taps are allowed only in surface raceways that are accessible after installation and have a removable cover. These splices may not occupy more than 75% of the cross-sectional area of the raceway at any point.

Splices for surface raceways that don't have removable covers are allowed only in junction boxes.

All sections of surface metal raceway must be mechanically joined so that the conductors won't be subjected to abrasion, and electrically connected so that they will all be properly grounded (Figure 10.1).

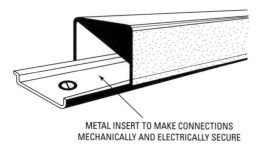

METAL INSERT TO MAKE CONNECTIONS
MECHANICALLY AND ELECTRICALLY SECURE

Figure 10.1   Connectors shall be electrically and mechanically secure.

Any nonmetallic items, such as covers, that are used with metallic surface raceways must be identified as suitable for that use.

Wire Fills

The sizes and numbers of conductors allowed in raceways are shown in the tables of Chapter 9, at the back of the 2002 NEC.

Surface Nonmetallic Raceways (*Article 388*)

Uses and Locations

May be used in dry locations.
   May *not* be used:

> Where the surface nonmetallic raceway might be subjected to severe physical damage.
>
> To enclose conductors that have a voltage of 300 V or higher between conductors, unless specifically listed for a higher voltage.
>
> In concealed locations.

In any hazardous locations, except when enclosing conductors in class I, division 2 locations that don't carry enough power to ignite specific atmospheres (see *Section 504.4(B)*).

To enclose conductors that have temperature limitations higher than the temperature limitations of the surface nonmetallic raceway.

Installation
Only unbroken pieces of surface metal raceway may be extended through floors, walls, or partitions.

When divided raceways (usually one section for power wiring, and one or more sections for data or communications wiring) are used, the positions of the sections relative to one another (such as data in the left-hand section and power in the right-hand section) must be maintained throughout the installation, to avoid confusion.

All sections of surface metal raceway must be mechanically joined so that the conductors won't be subjected to abrasion (Figure 10.2).

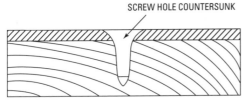

SCREW HOLE COUNTERSUNK

Figure 10.2 Holes for screws shall be countersunk to protect wire from damage.

Wire Fills
The sizes and numbers of conductors allowed in raceways are shown in the tables of Chapter 9, at the back of the 2002 NEC.

Multi-Outlet Assemblies (*Article 380*)

Uses and Locations
Multi-outlet assemblies may be used exposed, in dry locations.
   They may *not* be:

Concealed. The back and sides—but not the front—may be recessed into the building's finish.

Used where they may be subjected to physical damage.

Used to enclose conductors that have a voltage of 300 V or higher between conductors, unless the metal is at least 0.040 inches thick.

Used in areas where hazardous vapors are present.

Used in hoistways.

Used in hazardous locations, except when enclosing conductors in Class I, Division 2 locations that don't carry enough power to ignite specific atmospheres (see *Section 504.4(B)*).

Installation

It is allowable to install metal multi-outlet assemblies through partitions, as long as the cap covering all exposed portions can be removed and no outlet is located within the partition (Figure 10.3).

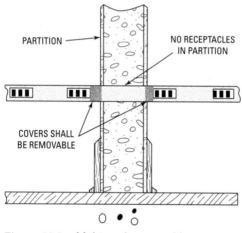

Figure 10.3   Multi-outlet assemblies may pass through dry partitions, provided there are no outlets in the partition and the covers are removable.

Underfloor Raceways (*Article 390*)

Uses and Locations

May be used when:

Underneath concrete floors.

Laid flush with finished floors (linoleum or equivalent) in offices.

May *not* be used:

Where they will be subjected to corrosive vapors.

In hazardous locations, except as allowed by *Section 504.20* and when enclosing conductors in Class I, Division 2 locations that don't carry enough power to ignite specific atmospheres (see *Section 504.4(B)(3)*).

Installation

Half-round or flat-top raceways that are 4 inches wide or less must be covered with at least ¾ inch of concrete or wood (Figure 10.4).

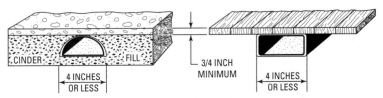

Figure 10.4 Minimum covering over 4-inch raceways.

Flat-top raceways between 4 and 8 inches wide and spaced at least 1 inch from other raceways must be covered with at least 1 inch of concrete. When these raceways are less than 1 inch apart, they must be covered with at least 1½ inches of concrete (Figure 10.5).

Only trench-type raceways with removable covers can be installed flush with the finished floor.

In offices, raceways no more than 4 inches wide with flat covers can be installed flush with the finished floor, but only if they are covered with at least ¹⁄₁₆ inch of linoleum or the equivalent. Up to three such raceways can be installed parallel to each other, but in such instances they must be joined together to form a rigid assembly.

All splices and taps must be made in junction boxes, *not* in the raceway itself. However, in trench ducts (with flat tops finished flush with the floor) that have removable covers that are accessible after installation, the splices may not occupy more than 75% of the cross-sectional area of the raceway.

Loop wiring (when conductors are looped at each outlet, not spliced, and put under a terminal screw with insulation removed but the conductor unbroken) is not considered to be a splice.

When any outlet is abandoned or removed, all conductors serving the outlet must be removed from the raceway. Looped

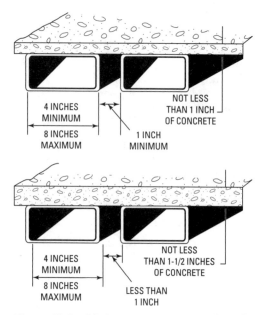

Figure 10.5 Minimum covering over 4- to 8-inch raceways.

conductors may *not* be reinsulated and put back in the raceway; they must be removed.

Underfloor raceways must be laid in straight lines, with no offsets, kicks, or bends. They must be firmly secured during concrete pouring, so that the straight-line alignment of the raceway won't be disturbed.

A marker must be placed near the end of every raceway, so that the last insert may be located easily.

All open ends of a raceway must be closed.

All junction boxes must be set level with the floor and sealed, so that no water may enter. Only metal boxes may be used with metal raceways, and they must be electrically bonded to the raceway.

All inserts must be set level with the floor and sealed, so that no concrete may enter during the pouring of the concrete. Only metal inserts may be used with metal raceways, and they must be electrically bonded to the raceway.

When fiber raceways are used, the inserts used with it must be secured to the raceway before concrete is poured. If inserts are to be added to fiber raceways after concrete is set, they must be screwed into the raceway. Proper tools must be used to do this, so that the tool does not protrude into the raceway and damage the conductors. Also, all debris must be removed from the raceway.

When underfloor raceways are to be connected to wall outlets or cabinets, one of the following methods must be used:

Rigid metal conduit.

Intermediate metal conduit.

Electrical metallic tubing.

Approved raceway fittings.

Underfloor raceway encased in concrete.

Flexible metal conduit.

If a metal underfloor raceway has provision for an equipment grounding conductor connection, connection between the raceway and wall outlets or cabinets may be made by any of the following methods, in addition to those mentioned above:

Rigid nonmetallic conduit.

Electrical nonmetallic tubing.

Liquid-tight flexible nonmetallic conduit.

Wire Fills

The sizes and numbers of conductors allowed in raceways are shown in the tables of Chapter 9, at the back of the 2002 NEC.

Cellular Metal Floor Raceways (*Article 374*)

Description

Cellular metal floor raceways use the open spaces of cellular floors to enclose wiring. Fittings are used to connect the cells (enclosed tubular spaces in the floor) and the area being served. These fittings extend through the concrete that is poured around the cells to make the finished floor. Header ducts are transverse raceways that provide access to the various cells (Figures 10.6, 10.7, 10.8).

Where Not Permitted

Cellular metal floor raceways may not be used:

Where corrosive vapors are present.

In hazardous locations, except as allowed by *Section 504.20* or when enclosing conductors in class I, division 2 locations

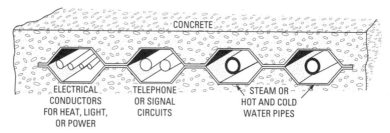

Figure 10.6 Cross-sectional view of cellular metal floor raceways used for installation of electrical and other systems.

Figure 10.7 Illustration of a cell.

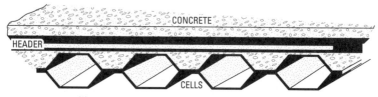

Figure 10.8 Illustration of a header and cells.

that don't carry enough power to ignite specific atmospheres (see *Section 504.4(B)(3)*).

In commercial garages, cellular metal floor raceways may be used only for supplying power to outlets below a floor, but never above a floor.

Installation

Conductors installed in these raceways must be no larger than 1/0 AWG, except when special permission is given by the authority having jurisdiction.

Splices are not permitted, except in header ducts or in junction boxes.

Loop wiring (when conductors are looped at each outlet, not spliced, and put under a terminal screw with insulation removed but the conductor unbroken) is not considered to be a splice.

When any outlet is abandoned or removed, all conductors serving the outlet must be removed from the raceway. Looped conductors may *not* be reinsulated and put back in the raceway; they must be removed.

A sufficient number of markers must be installed, so that cells can be located in the future.

All junction boxes must be set level with the floor and sealed, so that no concrete or water may enter. Only metal boxes may be used, and they must be electrically bonded to the raceway.

All inserts must be set level with the floor and sealed, so that no concrete may enter during the pouring of the floor. Only metal inserts may be used, and they must be electrically bonded to the raceway.

When cellular metal floor raceways are to be connected to wall outlets or cabinets, one of the following methods must be used:

Rigid metal conduit.

Intermediate metal conduit.

Electrical metallic tubing.

Approved raceway fittings.

Cellular metal floor raceway encased in concrete.

Flexible metal conduit.

If the raceway has provision for an equipment grounding conductor connection, connection between the raceway and wall outlets or cabinets may be made by any of the following methods, in addition to those mentioned above:

Rigid nonmetallic conduit.

Electrical nonmetallic tubing.

Liquid-tight flexible nonmetallic conduit.

All surfaces must be kept free of burrs or rough edges, which can damage the conductors. Smooth fittings or bushings must be used when conductors pass through the walls of the raceway.

Wire Fills

The sizes and numbers of conductors allowed in raceways are shown in the tables of Chapter 9, at the back of the 2002 NEC.

Cellular Concrete Floor Raceways (*Article 372*)

Description

Cellular concrete floor raceways use the open spaces of precast concrete floors to enclose wiring. Metal fittings are used to connect the cells (enclosed tubular spaces in the floor) and the area being served. Headers are transverse metal raceways that provide access to the various cells (Figure 10.9).

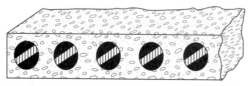

Figure 10.9   Cross-sectional view of a cellular concrete floor raceway.

Where Not Permitted

Cellular concrete floor raceways may not be used:

Where corrosive vapors are present.

In hazardous locations, except as permitted by *Section 504.20* or when enclosing conductors in class I, division 2 locations that don't carry enough power to ignite specific atmospheres (see *Section 504.4(B)(3)*).

In commercial garages, cellular concrete floor raceways may be used only for supplying power to outlets below a floor, but never above a floor.

Installation

Conductors installed in these raceways must be no larger than 1/0 AWG, except when special permission is given by the authority having jurisdiction.

Splices are not permitted, except in header ducts or in junction boxes.

Loop wiring (when conductors are looped at each outlet, not spliced, and put under a terminal screw with insulation removed but the conductor unbroken) is not considered to be a splice.

When any outlet is abandoned or removed, all conductors serving the outlet must be removed from the raceway.

Looped conductors may *not* be reinsulated and put back in the raceway; they must be removed.

A sufficient number of markers must be installed, so that cells can be located in the future.

All junction boxes must be set level with the floor and sealed, so that no concrete or water may enter. Only metal boxes may be used, and they must be electrically bonded to the raceway.

All inserts must be set level with the floor and sealed, so that no concrete or water may enter during the pouring of the floor. Only metal inserts may be used, and they must be terminated in grounding-type receptacles.

A separate grounding conductor must be used to connect the receptacles to a ground connection at the header.

If inserts are added to these raceways, or whenever the cell walls must be cut, proper tools must be used so that the tool does not protrude into the raceway and damage the conductors. Also, all debris must be removed from the raceway.

When cellular concrete floor raceways are to be connected to wall outlets or cabinets, metal raceways and approved fittings must be used.

Headers must be installed perpendicularly, in straight lines. The ends of the headers must be sealed with metal covers, and headers must be secured to the *top* of the precast concrete floors. Headers must be bonded to the distribution center enclosure and must be electrically continuous for their entire length.

Wire Fills

The sizes and numbers of conductors allowed in raceways are shown in the tables of Chapter 9, at the back of the 2002 NEC.

Metal Wireways (*Article 376*)

Uses and Locations

Metal wireways are allowed only in exposed locations. Note, however, that unbroken sections are permitted to pass transversely through walls, provided that the conductors are accessible on both sides of the wall. Metal wireways must be run in straight lines; the covers must be accessible; and no sharp or rough edges are permitted.

Metal wireways are *not* allowed:

Where they may be subjected to corrosive vapors.

Where they may be subjected to physical damage.

In hazardous locations, except where allowed by Section 501.4(b), 502.4(B), or 504.20 of the NEC.

In wet locations, except if listed for the purpose.

Installation

Splices or taps are allowed in accessible locations, but they may not occupy more than 75% of the wireway's cross-sectional area.

Wireways must have supports placed no more than 5 feet apart, except for pieces longer than 5 feet that are supported at each end or joint. Check to see if the wireway is specifically listed for support intervals. In no case can the supports be placed more than 10 feet apart.

Wireways that run vertically can have supports placed up to 15 feet apart, as long as there is no more than one joint between supports.

Unbroken sections of wireway are permitted to pass through walls (Figure 10.10).

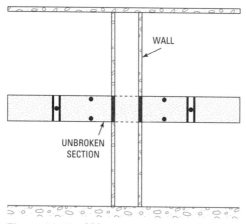

Figure 10.10   Wireways may extend transversely through a wall in unbroken lengths.

All open ends of wireways must be covered.

Wireways must be grounded (see *Article 250*).

Where grounding connections are made to wireways, any paint or coating must be removed from the area of the connection.

Wire Fills

Wireways are not allowed to contain more than 30 current-carrying conductors—neutral and grounding conductors or control wiring conductors are not counted—at any one cross-sectional area, except if the correction factors of *Section 310.15* are used.

The 30-wire limitation is not required in theaters and similar locations.

The total of the cross-sectional areas of all conductors in the wireway may not be greater than 20% of the cross-sectional area of the wireway.

Wireways for elevators are not subject to the preceding two requirements.

Auxiliary Gutters (*Article 366*)

Description
Auxiliary gutters are wireways that supplement wiring at meter centers, switchboards, etc.

Uses
Auxiliary gutters can enclose either conductors or bus bars.

Auxiliary gutters may *not* contain switches, appliances, overcurrent devices, or similar items.

Installation
Auxiliary gutters may not extend more than 30 feet from the equipment served, except when used for elevators.

Gutters must be supported at least every 5 feet.

Splices or taps are allowed in accessible locations, but they may not occupy more than 75% of the gutter's cross-sectional area.

All taps in gutters must have overcurrent protection (see *Section 240.21*).

They must also be marked, showing which circuit or equipment they supply.

Gutters installed in wet locations must be rain tight.

Bare conductors (such as buses) must have a clearance of at least 1 inch from all grounded surfaces and 2 inches from bare conductors of opposite polarity.

Wire Fills
Gutters are not allowed to contain more than 30 current-carrying conductors (neutral and grounding conductors or control wiring conductors are not counted) at any one cross-sectional area, except if the correction factors of *Section 310.15* are used.

The total of the cross-sectional areas of all conductors in the gutter may not be greater than 20% of the cross-sectional area of the gutter.

Gutters for elevators are not subject to the preceding two requirements.

Chapter 11

Busways

Busways (Busduct) (*Article 368*)

Uses and Locations
Busways are permitted:

> Where open and visible.
>
> Behind access panels, when:
>
>> No overcurrent devices are connected to the busway, except for individual devices.
>>
>> Only totally enclosed, nonventilating busway is used.
>>
>> The area behind the access panels is not used for air handling.
>>
>> The joints between the sections and fittings of the busway are accessible.

Busways may be installed in spaces used for air handling other than ducts and plenums if they are the totally enclosed, nonventilated type, with no provisions for plug-in connections.

Busways are *not* permitted:

> Where they might be subjected to corrosive vapors.
>
> Where they might be subject to physical damage.
>
> In hoistways.
>
> In hazardous locations, except in class I, division 2 and class II, division 2 locations where not enough power is present to ignite specific atmospheres. (See *Section 504.4(B)*.)
>
> Outdoors or in wet locations, except when identified as suitable for such use.

Installation
Busways must have supports placed no more than 5 feet apart, except if specifically marked otherwise by the manufacturer.

Unbroken lengths of busway are permitted to pass through dry walls (Figure 11.1).

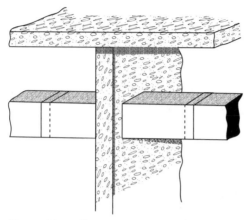

Figure 11.1 Busways may extend transversely through walls in unbroken lengths.

Totally enclosed nonventilated busways may pass through dry floors if it continues to at least 6 feet above the floor (Figure 11.2). All open ends of the busway must be closed.

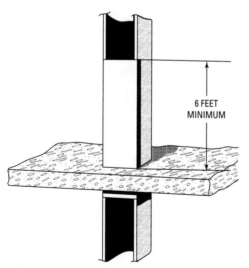

6 FEET
MINIMUM

Figure 11.2 Busways may extend vertically through floors (dry), if not ventilated, for a minimum height of 6 feet.

Branch and feeder circuits from busways may be run by any of the following methods:

Busway.

Rigid metal conduit.

Intermediate metal conduit.

Electrical metallic tubing.

Rigid nonmetallic conduit.

Electrical nonmetallic tubing.

Flexible metal conduit.

Metal surface raceway.

Metal-clad cable.

Cord assemblies that are approved for hard usage. (Allowed only for connection of portable equipment or equipment being interchanged, and strain reliefs must be installed on the cord. These cords must not extend more than 6 feet horizontally from the busway. There is no vertical limit if suitable means are used to take up the tension on the cord.)

In instances where the current rating of the busway does not exactly match standard ratings of overcurrent devices, the next larger size overcurrent protection device can be used.

All branch circuit or feeder taps from the busway must use plug-ins or connectors that have overcurrent devices in them. This is not required in only the following situations:

1. Where fixtures that have an overcurrent device mounted on them are mounted directly onto the busway.

2. Where an overcurrent device is part of a cord plug for cord-connected fixed or semi-fixed lighting fixtures.

3. For certain taps. See *Section 240.21*.

Busway runs over 600 volts must have vapor seals where they enter or leave a building. This is not required for forced-air busways.

Busway runs over 600 volts must have fire seals wherever they penetrate a floor, wall, or ceiling.

Drain plugs or drains must be installed at the low point of busways of 600 volts or more, in order to remove excess moisture.

When bus enclosures end at machines that are cooled by a flammable gas, they must use seal-off bushings or baffles, so that the gas can't build up in the enclosure.

Flexible connections or expansion joints must be used in long, straight runs.

All connections and terminations must be accessible after installation.

Cablebus (*Article 370*)

Uses and Locations

Cablebus is allowed for exposed work in any application for which its conductors are rated. But when used in outdoor, wet, or corrosive locations, it must be identified as being rated for such an application.

Cablebus is specifically permitted for systems in excess of 600 volts.

Cablebus is *not* permitted in hoistways or hazardous locations, unless specifically approved for use in these areas.

Installation

If the cablebus frame is properly bonded, it can be used as an equipment-grounding conductor.

The conductors used in cablebus must have a temperature rating of at least 75°C.

The current carrying capacity of conductors in cablebus is in accordance with *Tables 310.17* and *310.19*. This gives these conductors a free-air rating.

Conductors used in cablebus must be at least 1/0 AWG in size, and in cases must be the size and number of conductors for which the cablebus was designed.

Conductors must be supported at least every 3 feet in horizontal runs of cablebus, and every 1½ feet in vertical runs.

Conductors must be spaced at least one conductor diameter from each other in the cablebus.

If the current rating of the cablebus does not correspond with a standard rating, the next larger size can be used.

Cablebus must be supported every 12 feet or less, unless specifically designed for greater support spacings.

Cablebus can be routed transversely through walls and partitions (but NOT fire walls), as long as the portion through the wall is protected, continuous, and is not ventilated.

Cablebus can be routed vertically through dry floors and platforms, as long as the portion passing through the floor is totally enclosed where it passes through the floor, and for 6 feet above the floor. However, this is NOT allowed in areas where fire stops are required.

Where cablebus is routed through floors or platforms in wet locations, there must be curbs or similar means used to keep water from flowing through the opening in the floor or platform. Also, the cablebus must be totally enclosed where it passes through the floor, and for 6 feet above the floor. However, this is NOT allowed in areas where fire stops are required.

All sections of cablebus must be bonded together and grounded.

Chapter 12

Outlet and Pull Boxes

Outlet, Device, Pull, and Junction Boxes; Conduit Bodies, Fittings, and Manholes (*Article 314*)

Uses

Round boxes can't be used to terminate conduits that use lockouts or bushings to connect to boxes.

Nonmetallic boxes can be used only for nonmetallic conduits or cables, except when internal bonding means are used for all metal cables or conduits entering the box.

All metal boxes must be grounded.

Only boxes and fittings listed for use in wet locations can be used in wet areas. In addition, they must be arranged so that water won't enter or accumulate in the box.

Boxes in hazardous locations must conform to the specific requirements of the article governing the installation (see *Articles 500–517* of the NEC).

Unused openings in boxes must be filled. Openings in nonmetallic boxes may be filled with metal fillers, but the fillers must be recessed at least ¼ inch into the box.

Screws used for attaching a box can't be used for also attaching a device used in that box.

No more than a ¼-inch setback is allowed for boxes installed in noncombustible walls or ceilings.

Boxes set in combustible walls and ceilings must be set flush with the surface.

Plaster, drywall, or the like must be finished so that no gaps greater than ⅛ inch exist between the box and the plaster, etc.

All boxes must be securely mounted.

Boxes may *not* be mounted on support wires only.

Threaded enclosures without devices and occupying no more than 100 cubic inches of space can be mounted to two or more conduits without any other means of support. The conduits must be supported on two or more sides within 3 feet of the enclosure.

Conduit bodies can be supported by EMT or conduit, as long as their size is not larger than the largest size of conduit or EMT supporting them.

Threaded enclosures containing devices and occupying no more than 100 cubic inches of space can be mounted to two or more conduits, without any other means of support. The conduits must be supported within 18 inches of the enclosure.

Embedding an enclosure in concrete is considered adequate support.

Nonmetallic boxes can be supported by metal conduits under the same requirements as metal boxes, provided they are specifically listed for this use.

Boxes can be mounted from cord or cable pendants, but must use strain-relief connectors or some other suitable protection.

Every outlet box must have a suitable cover, canopy, or faceplate.

When canopy covers are used, any combustible material under the canopy (such as a wood ceiling) must be covered with noncombustible material (such as plaster).

Boxes used for lighting fixture outlets must be designed for that purpose.

Only floor boxes may be mounted in floors. Raised floors in show windows can use regular boxes, if the local authorities permit.

Boxes alone can't be used for mounting ceiling fans, unless specifically designed for that purpose.

Wire Fills

The number of wires permitted in a box is based on the area of the box, measured in cubic inches. Commercially available boxes are marked to show their cubic inch area. The areas required for different sizes of conductors are as follows:

Conductor Size (AWG)	Area Required (Cubic Inches)
No. 18	1.5
No. 16	1.75
No. 14	2.0
No. 12	2.25
No. 10	2.5
No. 8	3.0
No. 6	5.0

One wire must be deducted from the maximum box fill for any fixture stud, cable clamp, or hickey (Figure 12.1).

Two wires must be deducted from the maximum box fill for each mounting strap containing one or more wiring devices, such as receptacles or switches (Figure 12.2).

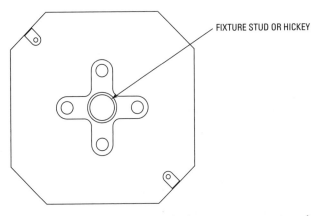

FIXTURE STUD OR HICKEY

Figure 12.1   A fixture stud or hickey counts as one conductor.

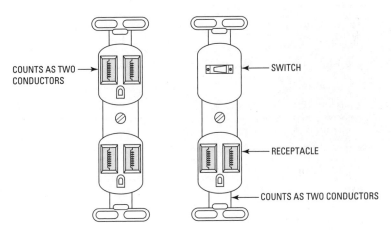

COUNTS AS TWO
CONDUCTORS

SWITCH

RECEPTACLE

COUNTS AS TWO CONDUCTORS

Figure 12.2   How to count devices in figuring fill.

When grounding conductors (one or more green wires) are present in a box, one wire must be deducted from the maximum fill. If a second set of grounding conductors (such as an isolated grounding system) is present, one additional wire also must be deducted from the maximum fill (Figure 12.3).

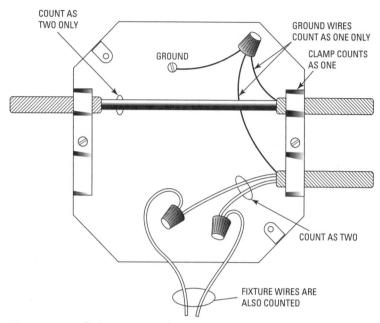

Figure 12.3 Only one grounding wire is counted.

A grounding conductor and/or up to four No. 14 fixture wires that enter a box under a fixture canopy can be omitted from these calculations. Additionally, branch-circuit conductors that go directly into a fixture canopy don't have to be counted as leaving the box.

When different sizes of conductors are in a box, the conductor deducted for a fixture stud or other device must be the largest size in the box; the conductors deducted for the device strap must be the largest size connected to the device strap.

A wire that runs straight through a box, and is not terminated in the box, counts as only one conductor (Figure 12.4).

The areas (in cubic inches) of plaster rings, raised covers, etc., may be included in the total area of the outlet box that may be filled.

Pull Boxes

When No. 4 AWG or larger conductors are contained in cable or conduits, associated pull or larger, associated pull or junction boxes must be sized as follows:

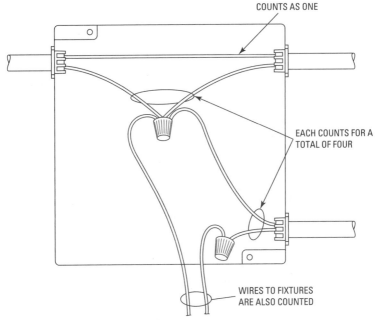

COUNTS AS ONE

EACH COUNTS FOR A
TOTAL OF FOUR

WIRES TO FIXTURES
ARE ALSO COUNTED

Figure 12.4 Which conductors to count in a junction box.

For straight runs, the length of the pull or junction box must be eight times the trade size of the largest raceway. If only cables are used (no raceways), the calculations must be made using the sizes of raceways that would be required if the conductors were in raceways rather than in cables (Figure 12.5).

For angle or U pulls, the distance between raceway entries in any wall of the box to the opposite wall must be at least six times the trade size of the largest raceway entering that wall. If only cables are used (no raceways), the calculations must be made using the sizes of raceways that would be required if the conductors were in raceways rather than in cables (Figure 12.6).

Any pull or junction boxes that have dimensions over 6 feet must have the conductors in the box cabled together or racked (Figure 12.7).

If permanent barriers are installed in a box (as would be the case where communication and power conductors share the same box), each section shall be considered a separate box.

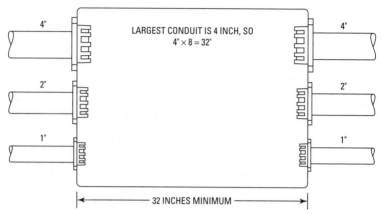

Figure 12.5 Calculation of pull boxes for use without splices or taps.

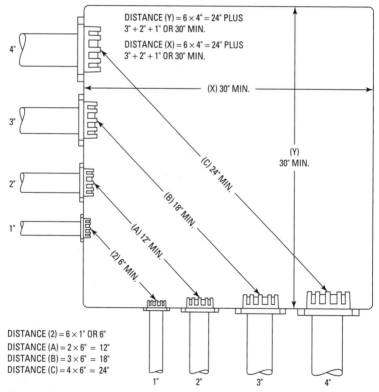

Figure 12.6 Junction box calculations.

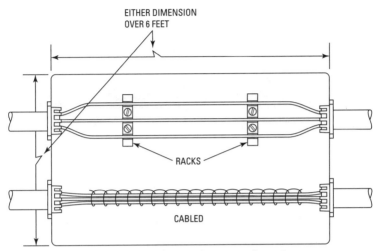

Figure 12.7   Cabling or racking conductors in large boxes.

All pull boxes, junction boxes, and conduit bodies must be installed so that the enclosed conductors are accessible following the installation.

Pull or junction boxes for systems of over 600 volts must have one or more removable sides.

Covers of boxes for systems of over 600 volts must be marked "DANGER—HIGH VOLTAGE—KEEP OUT."

When No. 4 or larger conductors enter a junction box or pull box, insulating bushings, approved fittings, or built-in hubs must be used to protect the conductors (Figure 12.8).

Cabinets and Cutout Boxes (*Article 312*)

Installation

Cabinets and cutout boxes (including meter fittings) that are installed in damp or wet locations must be spaced at least ¼ inch from the surface on which they are mounted. They must also be mounted so that moisture won't enter or accumulate.

No more than a ¼-inch setback is allowed for boxes installed in noncombustible walls or ceilings.

Boxes set in combustible walls or ceilings must be set flush with, or protruding from, the surface.

Boxes in hazardous locations must conform to the specific requirements governing installation of the article (see *Articles 500–517*).

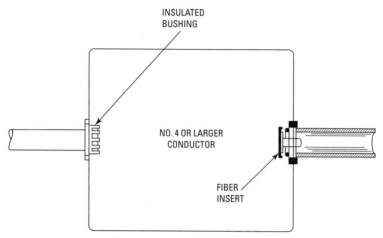

Figure 12.8 Use of insulation at bushings.

Unused openings in boxes must be filled. Openings in nonmetallic boxes may be filled with metal fillers, but the fillers must be recessed at least ¼ inch into the box.

When No. 4 or larger conductors enter a cabinet or cutout box, insulating bushings, approved fittings, or built-in hubs must be used to protect the conductors.

Cables with nonmetallic sheaths can enter cabinets through a conduit if they are supported within 8 inches of entering the cabinet, and if a fitting at the end of the conduit protects the cables from abrasion. Conduits containing cables entering through the top must be plugged to prevent debris from entering the box.

Chapter 13

Switches and Switchboards

Switches (*Article 404*)

Installation

Switching may *not* be done in grounded conductors.

Switches or circuit breakers can't disconnect grounded circuit conductors, except if all circuit conductors are simultaneously disconnected, or if the grounded conductor can't be disconnected until all other conductors are first disconnected.

Switches or circuit breakers installed in wet locations must be enclosed in weatherproof enclosures.

Switches (or circuit breakers used as switches) must be mounted so that the center of the operating handle is no more than 6 feet, 7 inches (2 meters) above the floor or working platform, except in the following cases:

1. Fused switches or circuit breakers on busway installations may be mounted at the same level as the busway; but provision must be made to operate the switch from the floor level.

2. Switches may be installed adjacent to motors, appliances, or other equipment above 6 feet, 7 inches, but these switches must be accessible by some portable method, such as stepladders.

3. Isolating switches that are operated by hooksticks may be mounted higher than 6 feet, 7 inches.

Switches must be installed so that the voltage between adjacent switches does not exceed 300 volts. If permanent barriers are installed between switches, they are not considered to be adjacent.

Switches that are mounted in nongrounded metal boxes must have faceplates made of nonconducting and noncombustible material.

Knife Switches

These switches are now obsolete, but they still exist in some places. See *Section 404.6* for further details.

Single-throw knife switches must either be installed so that gravity won't tend to close the switch, or be equipped with locking mechanisms (Figure 13.1).

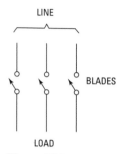

Figure 13.1
Position of knife switches.

When double-throw knife switches are mounted vertically, they must be equipped with locking mechanisms.

Knife switches must be installed so that the blades are deenergized when the switch is open, except when the load side of the switch is connected to circuits that by their very nature may provide a backfeed to the source of power. In these cases, a permanent sign must be placed adjacent to the switch, reading "WARNING—LOAD SIDE OF SWITCH MAY BE ENERGIZED BY BACKFEED."

Switchboards and Panelboards (*Article 408*)

Installation

Switchboards must have a clear area around them as specified in *Section 110.16* (see Chapter 1). For switchboards, this area must extend up to the structural ceiling, or 25 feet, whichever is lower. No piping, ducts, or other equipment (except sprinkler systems, which are specifically allowed) may be installed in these spaces. Ventilating equipment serving these areas, necessary control systems, and heavily protected equipment can be installed in these areas.

Three feet of space must exist between the top of a switchboard and a combustible ceiling, except for totally enclosed switchboards, or if a noncombustible shield is installed.

All switchboards must be grounded.

Panelboards installed in wet locations must have rain-tight enclosures and must be arranged so that no moisture will accumulate in the enclosure.

Any panelboards that are not of the dead-front type (which are now obsolete) must be installed in locations where they are accessible only to qualified personnel.

Chapter 14

Cords

Flexible Cords and Cables (*Article 400*)

See *Sections 400.4* and *400.5* for ampacity allowances for unusual cable types. This lengthy and rarely used section won't be included here.

Uses

Flexible cords and cables may be used:

As pendants.

For wiring fixtures.

To connect portable lamps or appliances. The cord or cable must be used with an attachment plug and fed from a receptacle outlet.

As elevator cables.

To wire cranes and hoists.

To connect stationary equipment that must be frequently interchanged. The cord or cable must be used with an attachment plug and fed from a receptacle outlet.

For connection of appliances that are identified for flexible cord connection. The cord or cable must be used with an attachment plug and fed from a receptacle outlet.

As data processing cables.

For the connection of moving parts.

As temporary wiring.

Flexible cords and cables may *not* be used:

As a substitute for fixed wiring in a structure.

In cases where it must be run through holes in walls, ceilings, or floors.

In cases where it must be run through doors, walls, or similar openings.

In attachment to building surfaces. One connection for a tension take-up device within 6 feet of the cord termination is allowed.

Behind walls, ceilings, or floors.

In raceways, except when specifically allowed under specific articles of the National Electrical Code.

Installation

Splices are allowed in cords only when they are required for the repair of existing cord installations. In these cases, the splices may be made with splicing devices, brazing, welding, or soldering. Soldered splices must be mechanically and electrically sound before the solder is applied. The completed splice must have the insulation, outer sheath, and usage characteristics of the cable being spliced.

Cords must be connected to devices or fittings so that no tension will be applied to the wire connection terminals. The recommended methods for doing this are:

Knotting the cord.

Wrapping the cord with electrical tape.

Using fittings that are designed for this purpose.

In show windows or showcases, the following types of cords can be used (except for connecting chain-hung lighting fixtures, portable lamps, or merchandise being displayed):

S, SE, SEO, SEOO, SO, SOO, SJ, SJE, SJEO, SJEOO, SJO, SJOO, SJT, SJTO, SJTOO, ST, STO, STOO, SEW, SEOW, SEOOW, SJEW, SJEOW, SJEOOW, SJOW, SJOOW, SJTW, SJTOW, SJTOOW, SOW, SOOW, STO, STOW, and STOOW.

Where flexible cords pass through holes or openings in boxes, covers, or other enclosures, they must be protected by bushings or fittings.

Portable Cables Over 600 Volts

All cables used on systems of more than 600 volts must contain conductors No. 8 AWG stranded copper or larger.

All cables operated at over 2000 volts must be shielded, and the shields must be grounded.

If connectors are used to connect lengths of cable, they must lock, and provisions must be made so that these connections won't be broken when the cables are energized. Tension must be eliminated at connections and terminations by appropriate methods.

Only permanent molded, vulcanized splices are allowed in these cables. Also, these splices must be accessible only to qualified personnel.

Chapter 15

Lighting Fixtures

Lighting Fixtures, Lampholders, and Lamps (*Article 410*)

Installation

All fixtures weighing more than 50 pounds must be supported independently from an outlet box.

Fixtures must be installed so that the wiring to the fixture can be inspected without being disconnected. Fixtures connected with flexible cords are excepted.

Fixtures can be used as raceways only in the following cases:

If the fixtures are marked as suitable for this use.

Fixtures that connect end-to-end, forming a continuous raceway (or fixtures connected by recognized wiring methods), can carry a 2-wire or multi-wire circuit through the fixtures. One additional 2-wire circuit may be installed if it feeds fixtures that are also connected to the previously mentioned 2-wire or multi-wire circuit. In such cases, these fixtures will have two different sources of power.

When wires are installed within 3 inches of a ballast, they must have at least a 90° C rating, such as type THHW, THHN, or XHHW.

Parts of suspended ceilings used to support lighting fixtures must be secured to each other, and to the structural ceiling.

Recessed fixtures must have at least a 3-inch clearance from thermal insulation and a ½-inch clearance from combustible material, unless the fixture is specifically listed for use with less clearance.

Remote fixture ballasts can't be installed in contact with combustible materials.

Whenever raceway fittings are used to support fixtures, they must be able to support the entire weight of the fixture and any associated lamps, shades, etc.

Fixtures installed in wet locations must be arranged so that water won't enter into or accumulate in them. Fixtures installed in wet locations and damp locations must be listed and marked as suitable for these locations.

Fixture installations in concrete or masonry in contact with the earth must be considered wet locations.

Fixture installations in areas such as basements, cold-storage warehouses, and some barns must be considered damp locations.

Fixtures installed in corrosive locations must be suitable for the purpose.

When fixtures are to be installed in cooking hoods in nonresidential locations, the following requirements must be met:

> The fixtures must be vapor-tight, and the diffusers must be thermal shock-resistant.
>
> The fixtures must be identified as suitable for the purpose, and the materials being used must be suitable for the heat encountered in the area where they are installed.
>
> The exposed parts of the fixture within the hood must be corrosion-resistant, and must have smooth edges so that grease won't accumulate.
>
> The wiring supplying the hood (including cable or raceway) must not be exposed within the hood.

No pendant, hanging, or cord-hung fixtures can be installed within 3 feet horizontally from, or within 8 feet above, any bathtub rim.

All fixtures must be installed so that no combustible materials will be subjected to temperatures over 90° C (194° F).

When lampholders are to be installed above highly combustible materials, they must not be of the switched type (with pull-chain); and if individual switches are not installed for each lampholder, they must be at least 8 feet above the floor and be protected or located so that the lamps (bulbs) can't be easily removed or damaged.

In show windows, only chain-hung fixtures can be externally wired (Figures 15.1, 15.2).

When fixtures are to be installed in coves, the coves must have enough space to properly install and maintain the fixtures.

Branch circuits are not permitted to pass through incandescent fixtures unless the fixtures are marked as suitable for that use.

All outlet boxes used for lighting must have a suitable canopy or cover.

Any combustible wall surfaces (such as wood ceilings) under fixture canopies must be covered with a noncombustible material (such as plaster).

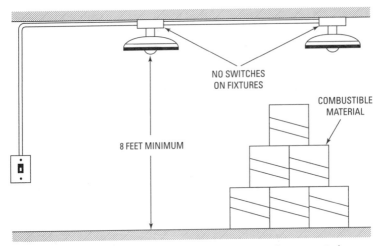

Figure 15.1   Fixtures over combustibles must not have a switch as part of the fixture but must be switched elsewhere.

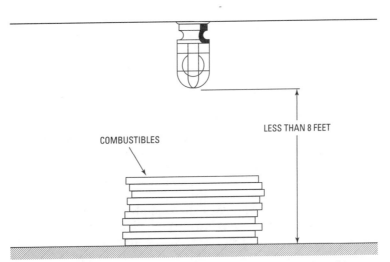

Figure 15.2   Guarding and switching of fixtures over combustibles when fixture is less than 8 feet from the floor.

When electric-discharge lighting fixtures are supported independently of outlet boxes, they can be connected with any of the following:

> Metal raceways.
>
> Nonmetallic raceways.
>
> MC cable.
>
> AC cable.
>
> MI cable.
>
> Nonmetallic-sheathed cable.
>
> Flexible cords—only if the cord is entirely visible, not subject to strain or damage, and is terminated in a grounded attachment plug or busway plug.

All fixtures, except those having no exposed conductive parts, must be grounded.

Whenever a grounded circuit conductor (neutral wire) connects to a screwshell lampholder, it must connect to the screwshell rather than the tab inside the lampholder.

Only necessary splices and connections are allowed to be made within fixtures.

No splices or taps can be made in fixture stems, arms, or similar locations.

Up to six cord-connected showcases (showcases connected by cord and locking connectors) can be connected to a permanent receptacle outlet. The cord conductors must be at least as large as the branch-circuit conductors, a grounding receptacle must be used, and no equipment outside of the showcases may be connected to the cord-supplied circuits. Also, the cases must be no more than 2 inches apart, and the first showcase must be within 12 inches from the receptacle. The wiring can't be subject to damage, and if a free lead is installed at the end of a run, it must be a female cord end and can't extend beyond the end of the case.

For paired fixtures with one ballast supplying lamps in both fixtures, the wiring from a ballast to a remote fixture can be run up to 25 feet in $\frac{3}{8}$-inch flexible metal conduit. Fixture wire carrying line voltage that supplies only the ballast in one of the paired fixtures may also be installed in this raceway.

Fixtures in Clothes Closets

The installation of various types of lighting fixtures in clothes closets is allowed according to how close the fixtures are to what is defined as *storage space*. Storage space is:

> The area within 24 inches from back and walls of the closet, and from the floor up to the highest clothes-hanging rod or 6 feet, whichever is higher. Above the highest rod or 6 feet, whichever is higher, the area within 12 inches of the back or walls is considered storage space. If shelves are wider than 12 inches, the area above the shelves (whatever the width of the shelves) must be considered a storage area.

The following types of fixtures are permitted in closets:

> Surface-mounted or recessed incandescent fixtures that have a completely enclosed bulb.
>
> Surface-mounted or recessed fluorescent fixtures.

The following types are *not* permitted:

> Incandescent fixtures with open or partially exposed lamps may not be used in closets.
>
> Pendant fixtures and lampholders (pull-chain or keyless fixtures) are also prohibited.

Fixtures in clothes closets must be installed as follows:

> Surface-mounted incandescent fixtures can be mounted on the wall above a door, as long as there is a distance of at least 12 inches between the bulb and the storage area.
>
> Surface-mounted fluorescent fixtures can be mounted on the wall over a door, as long as there is a distance of at least 6 inches between the bulb(s) and the storage area.
>
> Recessed incandescent fixtures that completely enclose the lamp (the fresnel type may be used, but the baffle type can't) can be installed in the wall or ceiling, as long as there is a distance of at least 6 inches between the surface of the fixture and the storage area.
>
> Recessed fluorescent fixtures can be mounted in the wall or ceiling, as long as there is a distance of at least 6 inches between the surface of the fixtures and the storage area.

Track Lighting

Lighting track must be permanently attached and connected to a branch circuit.

The branch circuit that supplies a lighting track must have a rating equal to or greater than that of the lighting track.

The connected load on a lighting track can't be greater than the rating of the branch circuit that feeds the track. This can easily happen if too many track fixtures (heads) are connected to the track.

Track lighting can't be installed in the following locations:

In damp or wet locations.

In storage battery rooms.

In corrosive areas.

In hazardous locations.

In concealed areas.

Through walls or partitions.

Within 5 feet above the finished floor, except when protected.

Single sections of track must have at least two supports. Longer runs should be supported at least every 4 feet.

Track lighting must be securely grounded throughout its entire length.

Chapter 16

Receptacles

Receptacles, Cord Connectors, and Attachment Plugs (*Article 406*)

Receptacles

Receptacles used for the connection of portable cords must be rated no less than 15 amps/125 volts or 15 amps/250 volts.

Any receptacles rated 20 amps or less that are to be connected to aluminum wiring must be marked CO/ALR.

All metal faceplates must be grounded. This is almost always done automatically, simply by screwing the faceplate onto the grounded yoke of the wiring device.

Faceplates must seat against the mounting surface (the wall surface) and completely cover any opening.

If the outlet box containing a receptacle is set back from the wall surface, the yoke of the receptacle must be held rigidly against the surface of the wall. In all cases, the yoke of a receptacle must be rigidly mounted and grounded.

Receptacles installed in damp or protected outdoor locations must be equipped with covers that are weatherproof when closed (when a plug is not inserted). Protected locations are considered to be areas that are under porches, canopies, and the like, where receptacles won't be directly subjected to beating rain or runoff water.

Receptacles installed in wet locations must be equipped with covers that protect the receptacle with the plug inserted, unless a self-closing weatherproof cover is used and only portable tools or equipment are temporarily connected.

Outdoor receptacles are to be located so that water is not likely to touch the outlet cover.

Grounding terminals on receptacles must *never* be used for anything but grounding.

Floor receptacles must be installed so that floor-cleaning equipment can be used without damaging the receptacles.

Attachment Plugs and Cord Connectors

Grounding attachment plugs can be used *only* where an equipment ground is present.

Grounding terminals on attachment plugs may *never* be used for anything but grounding.

Chapter 17

Appliances

Appliances (*Article 422*)

Branch Circuits

A branch circuit that supplies an appliance must not be rated at less than the nameplate rating of the appliance it serves.

See Chapter 2 for requirements pertaining to branch circuits that supply more than one appliance.

Central heating equipment must be supplied by an individual branch circuit. Auxiliary equipment such as humidifiers, air filters, etc. can be connected to this circuit also.

Branch circuits that supply storage-type water heaters must be rated at least 125% of the nameplate rating of the water heater.

Installation

All appliances must be installed in an approved manner.

All exposed noncurrent-carrying metal parts of appliances must be grounded.

Flexible-cord connections are allowed in the following circumstances:

> To connect appliances that must be frequently interchanged or that would be adversely affected by vibration.
>
> Where ready removal for maintenance or repair is required. In this case, the appliance must be identified as suitable for such use.

Kitchen disposers in dwellings can be equipped with cord connections if:

> The cord length is between 18 and 36 inches.
>
> A grounding plug is used, except if the unit is double-insulated and marked accordingly.
>
> The receptacle is located so that the cord won't receive physical damage.
>
> The receptacle is accessible.
>
> Cords listed for such installations are used.

Built-in dishwashers and trash compactors in dwellings can be equipped with cord connections if:

The cord length is between 3 and 4 feet.

A grounding plug is used, except if the unit is double-insulated and marked accordingly.

The receptacle is located so that the cord won't receive physical damage.

The receptacle is accessible.

Cords listed for such installations are used.

Electrically heated appliances must be located so that protection is provided between the appliance and any combustible materials.

Wall-mounted ovens and counter-mounted cooking units (note that these are *not* electric ranges, but wall ovens and cooktops) can be either permanently connected (hard-wired) or cord-and-plug connected.

If cord-and-plug connected, the plug and receptacle combination can't be considered a disconnecting means, and must be suitable for the temperatures it will encounter.

All appliances must have a means of disconnection. Any appliances fed by more than one source must have a disconnecting means for each source, and these disconnecting means must be grouped together and identified. The various disconnecting means allowed are as follows:

For permanently connected appliances of 300 watts (watts are also expressed as VA) or less, the branch circuit overcurrent protection device (fuse or circuit breaker) is considered the disconnecting means.

For permanently connected appliances of greater than 300 watts, the branch circuit switch or breaker can serve as the disconnecting means if it is within sight of the appliance or can be locked in the open position.

For cord-connected appliances, a receptacle and plug or accessible connector can serve as the disconnecting means.

A cord-and-receptacle arrangement can be used as the disconnecting means for an electric range, if the connection is accessible from the front of the range by removing a drawer.

An on/off switch in an appliance may be used as the disconnecting means, except in multi-family dwellings where the branch-circuit overcurrent device feeding the appliance is not on the same floor as the dwelling unit.

Switches or circuit breakers used as disconnecting means must be the indicating type—with "ON" and "OFF" marked on them.

Fixed Electric Space-Heating Equipment (*Article 424*)

Branch Circuits
Branch circuits must be rated at least 125% of all heating and motor loads.

If the rating of the circuit does not correspond to a standard overcurrent protection device rating, the next higher size can be used.

Thermostats that are rated for continuous use at 100% of the rated load don't have to be sized to 125% of the circuit rating.

Installation
Heaters to be installed in damp or wet locations must be specifically listed for such use, and must be installed so that water won't enter or accumulate in the unit.

Fixed electric space-heating equipment may not be installed where it can be subjected to physical damage.

These heaters must be installed so that adequate space is provided between the units and any combustible materials, except if specifically listed as usable in contact with combustible materials.

All noncurrent-carrying metal parts of fixed electric space heaters must be grounded.

An on/off switch in a fixed electric space heater may be used as the disconnecting means, except in multi-family dwellings where the branch-circuit overcurrent device feeding the appliance is not on the same floor as the dwelling unit.

A thermostat will be considered as both a controller and disconnecting means as long as it has a clearly marked "off" position, it opens all ungrounded conductors in the "off" position, it is within sight of the heater, and the system is designed so that it can't be energized when the thermostat is in the "off" position.

Switches or circuit breakers used as disconnecting means must be the indicating type—with "ON" and "OFF" marked on them.

Heaters having motor-compressors come under the authority of *Article 440*. (See Chapter 19.)

Electric Space-Heating Cables

Wiring above heated ceilings must not be more than 2 inches above the ceilings. The wires must have their ampacity calculated according to a 50° C ambient temperature. The multipliers are shown in *Tables 310.16, 310.17, 310.18,* and *310.19.*

Heating cables can't extend outside of the room of their origination.

Heating cables can't be installed above walls or partitions that extend to the ceiling (except for isolated, single embedded cables), in closets, or above cabinets that are closer to the ceiling than their depth (Figure 17.1).

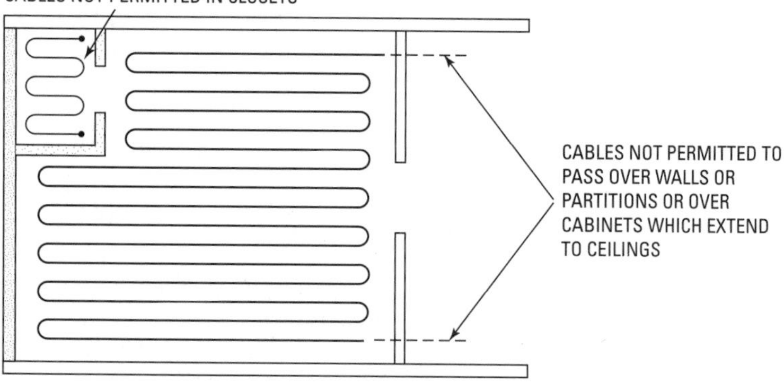

Figure 17.1 Illustration showing where heating cables may and may not be installed in ceilings.

Special low-heat cables for humidity control are allowed to be installed in closets, but only in areas that are not over shelves.

Heating cables must be at least 8 inches from outlet boxes, and at least 2 inches from recessed fixtures, trims, or vents.

Embedded heating cables may be spliced only when necessary.

Heating cables can't be installed in walls, except to go from one ceiling level to another.

Heating cables can be installed only on fire-resistant materials. Any exposed metal lath must be covered with plaster before the heating cables are installed.

Runs of cables rated 2.75 watts per foot or less must be at least 1½ inches apart (center to center).

All of the heating cables and at least 3 inches of the nonheating leads must be embedded in plaster or dry board.

Cables must be secured at least every 16 inches, except for cables specifically identified for different support spacings, but never more than every 6 feet.

On drywall ceilings, the heating cables must be installed to the joists, and at least 2½ inches on center. The cables may cross joists only when necessitated by obstructions or at the ends of a room.

The nonheating leads for heating cables can be brought from the junction box to the ceiling by one of the following methods:

As single conductors in raceways.

As single or multiconductor UF, NMC, or MI cables.

Excess leads of heating cables shouldn't be cut, but coiled in the ceiling, with only enough extended to the junction box so that 6 or 8 inches of lead is free within the box.

In concrete or poured masonry floors, heating cables can't exceed 16.5 watts per foot, and may not be placed closer than 1 inch apart on centers.

The cables must be secured by nonmetallic means while the concrete is being poured.

Cables may not be installed across expansion joints unless protected.

Heating cables (except grounded metal-clad cables) must be kept separate from metal in the concrete or masonry.

When leads leave the floor, they must be protected with bushings and use one of the following methods:

Rigid metal conduit

Intermediate metal conduit

Rigid nonmetallic conduit

Electrical metallic tubing

Duct Heaters

Heaters used in ducts must be identified as suitable for such use. Units to be used within 4 feet of heat pumps need additional markings for such use on both the heater and the heat pump.

Caution must be taken to ensure air flow through the heater according to the manufacturer's instructions.

Duct heaters must be accessible after installation and be installed according to the manufacturer's instructions.

A disconnecting means must be installed within sight of the heater's controller.

Ice and Snow Melting Equipment (*Article 426*)

Installation

Cables may be installed no closer than 1 inch on center.

Cables must be installed on an asphalt or masonry base at least 2 inches thick, and must be covered by at least 1½ inches of asphalt or masonry. Other investigated methods may also be used.

Cables must be secured while the masonry or asphalt is being applied.

When cables are installed across expansion joints, provision must be made for expansion and contraction.

Nonheating leads in masonry or asphalt must have a ground sheath or braid, or else have additional protection.

Nonheating leads enclosed in raceways must have between 1 and 6 inches of length. Bushings must be used to protect the cables as they leave the raceways. Rigid metal conduit, intermediate metal conduit, electrical metallic tubing, or other raceways may be used.

All noncurrent-carrying metal parts must be grounded.

Equipment on branch circuits that supply fixed outdoor electric de-icing and snow melting must have ground-fault protection.

Chapter 18

Motors and Controllers

Motor Circuits, Controllers (*Article 430*)

Adjustable-Speed Drive Systems

The size of branch circuits or feeders to adjustable-speed drive equipment must be based on the rated current input to the equipment.

If overload protection is accomplished by the system controller, no further overload protection is required.

The disconnecting means for adjustable-speed drive systems can be installed on the incoming line, and must be rated at least 115% of the conversion unit's input current.

Part-Winding Motors

If separate overload devices are used with standard part-winding motors, each half of the windings must be separately protected, at one-half the trip current specified for a conventional motor of the same horsepower rating.

Each winding must have separate branch-circuit, short-circuit, and ground-fault protection, at no more than one-half the level required for a conventional motor of the same horsepower rating.

A single device (which has the one-half rating) can be used for both windings if it will allow the motor to start.

If a single time delay fuse device is used for both windings, it can be rated at no more than 150% of the motor's full-load current.

Motor Ratings and Ampacity Ratings

Every motor is considered to be continuous-duty unless the characteristics of the equipment it drives ensure that the motor can't operate under load continuously.

Except for torque motors and AC adjustable-voltage motors, the current rating of motors (this rating is used to determine conductor ampacities, switch ratings, and branch-circuit ratings) must be taken from *Tables 430.147, 148, 149,* and *150* of the NEC. These values may *not* be taken from a motor's nameplate rating, except for shaded-pole and permanent-split-capacitor fan or blower motors, which are rated according to their nameplates.

Separate overload protection for motors is to be taken from the motor's nameplate rating.

Multispeed motors must have the conductors to the line side of the controller rated according to the highest full-load current shown on the motor's nameplate (so long as each winding has its own overload protection, sized according to its own full-load current rating). The conductors between the controller and the motor are to be based on the current for the winding, which is supplied by the various conductors.

Torque Motors
The rated current for torque motors must be the locked-rotor current. This nameplate current must be used in determining branch-circuit ampacity, overload, and ground-fault protection.

AC Adjustable-Voltage Motors
The ampacity of branch circuits, switches, short-circuit, and ground-fault protection for these motors must be based on the full-load current shown on the motor's nameplate. If no nameplate is present, these ratings must be calculated as no less than 150% of the values shown in *Tables 430.149* and *430.150*.

Motor Locations
In general, motors must be installed so that adequate ventilation is provided and so that maintenance operations can be performed.

Open motors with commutators or collector rings must be located so that sparks from the motors can't reach combustible materials. However, this does not prohibit these motors on wooden floors.

Suitable enclosed motors must be used in areas where significant amounts of dust are present.

Motor Circuit Conductors
Branch-circuit conductors that supply single motors must have an ampacity of at least 125% of the motor's full-load current rating, as taken from *Tables 430.147* through *430.150*.

Motors used only for short cycles can have their branch-circuit ampacities reduced according to *Table 430.22(E)*.

DC motors fed by single-phase rectifiers must have the ampacity of their conductors rated at 190% of full-load current for half-wave systems, and 150% of full-load current for full-wave systems.

For phase converters, the single-phase conductors that supply the converter must have an ampacity of at least 2.16 times the full-load current of the motor or load being served. This assumes that the voltages are equal. If they are not, the calculated current must be multiplied by the result of output voltage divided by input voltage.

Conductors connecting secondaries of continuous-duty wound-rotor motors to their controllers must have an ampacity of at least 125% of the full-load secondary current.

When a resistor is installed separate from a controller, the ampacity of the conductors between the controller and the resistor must be sized according to *Table 430.23(C)*.

Conductors Supplying Several Motors or Phase Converters

Conductors that supply two or more motors must have an ampacity of no less than the total of the full-load currents of all motors being served, plus 25% of the highest-rated motor in the group. If interlock circuitry guarantees that all motors can't be operated at the same time, the calculations can be made based on the largest group of motors that can be operated at any time.

Several motors can be connected to the same branch circuit if the following requirements are met:

Motors of one horsepower or less can be installed on general purpose branch circuits without overload protection (assuming all other requirements are met).

The full-load current can't be more than 6 amperes.

The branch-circuit protective device marked on any controller is not exceeded.

The conductors mentioned above must be provided with a protective device rated no greater than the highest rating of the protective device of any motor in the group, *plus* the sum of the full-load currents of the other motors.

Where heavy-capacity feeders are to be installed for future expansions, the rating of the feeder protective devices can be based on the ampacity of the feeder conductors.

Phase converters must have an ampacity of 1.73 times the full-load current rating of all motors being served, plus 25% of the highest-rated motor in the group. This assumes that the voltages are equal. If they are not, the calculated current must be multiplied by the result of output voltage divided by input voltage. If the ampere rating of the 3-phase output conductors is less than 58% of the rating of the single-phase input current ampacity, separate overcurrent protection must be provided within 10 feet of the phase converter.

Conductors Supplying Motors and Other Loads

Conductors that supply motors and other loads must have their motor loads computed as specified above; other loads computed according to their specific code requirements; and the two loads added together.

If taps are to be made from feeder conductors, they must terminate into a branch-circuit protective device and must:

> Have the same ampacity as the feeder conductors *or* be enclosed by a raceway or in a controller.
>
> Be no longer than 10 feet *or* have an ampacity of at least one-third of the feeder ampacity.
>
> Be protected.
>
> Be no longer than 25 feet.

In high-bay manufacturing buildings (which are more than 35 feet from floor to ceiling, measured at the walls), taps longer than 25 feet are permitted. In these cases:

> The tap conductors must have an ampacity of at least one-third that of the feeder conductors.
>
> The tap conductors must terminate in an appropriate circuit breaker or set of fuses.
>
> The tap conductors must be protected from damage and installed in a raceway.
>
> The tap conductors must be continuous, with no splices.
>
> The minimum size of tap conductors is No. 6 AWG copper, or No. 4 AWG aluminum.
>
> The tap conductors can't penetrate floors, walls, or ceilings.
>
> The tap conductors may be run no more than 25 feet horizontally, and no more than 100 feet overall.

Feeders that supply motors and lighting loads must be sized to carry the entire lighting load, plus the motor load.

Overload Protection

Overload protection is not required where it might increase or cause a hazard, as would be the case if overload protection were used on fire pumps.

Continuous-duty motors of more than one horsepower must have overload protection. This protection may be in one of the following forms:

> An overload device that responds to motor current. These units must be set to trip at 115% of the motor's full-load

current (based on the nameplate rating). Motors with a service factor of at least 1.15 or with a marked temperature rise of no more than 40° C can have their overloads set to trip at 125% of full-load current.

By any of several methods requiring overload protection built into motors. This would be done by the motor manufacturer, not by the installer.

Motors of one horsepower or less that are nonpermanently installed, manually started, and within sight of their controller are considered to be protected from overload by their branch-circuit protective device. They may be installed on 120-volt circuits of up to 20 amps.

Motors of one horsepower or less that are permanently installed, automatically started, or not within sight of their controllers may be protected from overloads by one of the following methods:

An overload device that responds to motor current. These units must be set to trip at 115% of the motor's full-load current (based on the nameplate rating). Motors with a service factor of at least 1.15 or with a marked temperature rise of no more than 40° C can have their overloads set to trip at 125% of full-load current.

By any of several methods requiring overload protection built into motors. This would be done by the motor manufacturer, not by the installer.

Motors that have enough impedance to ensure that overheating is not a threat can be protected by their branch-circuit protection method only.

Wound-rotor secondaries are considered to be protected from overload by the motor overload protection.

Intermittent-duty motors can be protected from overload by their branch-circuit protective devices only, as long as the rating of the branch-circuit protective device does not exceed the rating specified in *Table 430.152*.

In cases where normal overload protection is too low to allow the motor to start, it may be increased to 130% of the motor's full-load-rated current. Motors with a service factor of at least 1.15 or with a marked temperature rise of no more than 40° C can have their overloads set to trip at 140% of full-load current.

Manually-started motors are allowed to have their overload protection devices momentarily cut out of the circuit during the starting period. The design of the cutout mechanism must ensure that it won't allow the overload protective devices to remain cut out of the circuit.

If fuses are used as overload protection, they must be installed in all ungrounded motor conductors, and also in the grounded conductor for 3-phase, 3-wire systems that have one grounded wire (corner-ground delta systems).

If trip coils, relays, or thermal cutouts are used as overload protective devices, they must be installed according to *Table 430.37*. The requirements for standard motor types are as follows:

3-phase ac motors—one overload device must be placed in each phase.

Single-phase ac or dc, 1 wire grounded—one overload device in the ungrounded conductor.

Single-phase ac or dc, ungrounded—one overload device in either conductor.

Single-phase ac or dc, 3 wires, grounded neutral—one overload device in either ungrounded conductor.

In general, overload protective devices should open enough ungrounded conductors to stop the operation of the motor.

When motors are installed on general-purpose branch circuits, their overload protection must be as follows:

Motors of more than one horsepower can be installed on general-purpose branch circuits only when their full-load current is less than 6 amperes; they have overload protection; the branch-circuit protective device marked on any controller is not exceeded; and the overload device is approved for group installation.

Motors of one horsepower or less can be installed on general-purpose branch circuits without overload protection (assuming it complies with the other requirements mentioned above); the full-load current can't exceed 6 amperes; and the branch-circuit protective device marked on any controller is not exceeded.

When a motor is cord-and-plug-connected, the rating of the plug and receptacle may not be greater than 15 amperes at 125 volts or 10 amperes at 250 volts. If the motor is more than one horsepower, the overload protection must be built into the motor. The branch circuit must be rated according to the rating of the cord and plug.

The branch and overload protections must have enough time delay to allow the motor to start.

Overload protection devices that can restart a motor automatically after tripping are not allowed unless they are approved for use with a specific motor. This is *never* allowed if it can possibly cause personal injury.

In cases where the instant shutdown of an overloaded motor would be dangerous to persons (as could be the case in a variety of industrial settings), a supervised alarm may be used and an orderly—rather than instant—shutdown can be done.

Short-Circuit and Ground-Fault Protection

Short-circuit and ground-fault protective devices must be capable of carrying the starting current of the motors they protect.

In general, protection devices must have a rating of no less than the values given in *Table 430.52*. When these values don't correspond with the standard ratings of overcurrent protection devices, the next higher setting can be used instead.

If the rating given in *Table 430.52* is not sufficient to allow for the motor's starting current, the following methods can be used:

A non-time delay fuse of 600 amps or less can be increased enough to handle the starting current, but *not* to more than 400% of the motor's full-load current.

A time delay fuse can be increased enough to handle the starting current, but *not* to more than 225% of the motor's full-load current.

The rating of an inverse-time circuit breaker can be increased, but not to more than 400% of full-load currents that are 100 amps or less, or 300% of full-load currents that are over 100 amps.

The rating of an instantaneous-trip circuit breaker can be increased, but not to more than 1300% of full-load current.

Fuses rated between 601 and 6000 amps can be increased, but not to more than 300% of the rated full-load current.

Instantaneous-trip circuit breakers can be used as protective devices, but only if they are adjustable and part of a listed combination controller that has overload, short-circuit, and ground-fault protection in each conductor.

A motor short-circuit protector can be used as a protective device, but only when it is part of a listed combination controller that has overload, short-circuit, and ground-fault protection in each conductor and does not operate at more than 1300% of full-load current.

For multispeed motors, a single short-circuit and ground-fault protective device can protect two motor windings, as long as the rating of the protective device is not greater than the highest possible rating for the smallest winding. Multipliers of *Table 430.52* are included in the smallest winding's rating.

A single short-circuit and ground-fault protective device can be used for multispeed motors, sized according to the full-load current of the highest-rated winding. But each winding must have its own overload protection, which must be sized according to each winding. Also, the branch-circuit conductors feeding each winding must be sized according to the full-load current of the highest winding.

If branch-circuit and ground-fault protection ratings are shown on motors or controllers, they must be followed, even if they are lower than National Electrical Code requirements.

Fuses can be used instead of the devices mentioned in *Table 430.52* for adjustable-speed drive systems, as long as a marking for replacement fuses is provided next to the fuse holders.

For torque motors, the branch-circuit protection must be equal to the full-load current of the motor. If the full-load current is 800 amps or less and the rating does not correspond to a standard overcurrent protective device rating, the next higher rating can be used. If the full-load current is more than 800 amps and different from a standard overcurrent device rating, the next *lower* rating must be used.

If the smallest motor on a circuit has adequate branch-circuit protection, additional loads or motors can be added to the circuit. But each motor must have overload protection, and there must be assurance that the branch-circuit protective device won't open under the most stressful of normal conditions.

Two or more motors (each motor having its own overload protection) or other loads may also be connected to a single circuit in the following cases:

> When the overload devices are factory-installed, and the branch-circuit, short-circuit, and ground-fault protection are part of the factory assembly, or are specified on the equipment.

> When the branch-circuit protective device, motor controller, and overload devices are separate field-installed assemblies,

are listed for this use, and are provided with instructions from the manufacturer.

When all overload devices and motors are marked as suitable for group installation and are marked with a maximum rating for fuses and/or circuit breakers. Each circuit breaker must be of the inverse-time type, and must be listed for group installation.

When the branch circuit is protected with an inverse-time circuit breaker that is rated for the highest-rated motor and all other loads (including motor loads) that are connected to the circuit.

When branch-circuit fuses or inverse-time circuit breakers are not larger than allowed for the overload relay that protects the smallest motor in the group (see *Section 430.40).*

For group installations as described above, taps to single motors don't need branch-circuit protection in any of the following cases:

The conductors to the motor have an ampacity equal to or greater than the branch-circuit conductors.

The conductors to the motor are no longer than 25 feet, are protected, and have an ampacity at least one-third as great as the branch-circuit conductors.

Motor Control Circuits

Motor control circuits that are tapped from the load side of a motor's branch-circuit device and control the motor's operation are not considered branch circuits. They can be protected by either a supplementary or a branch-circuit protective device. Control circuits that are not tapped this way are considered signaling circuits and must be protected accordingly (see *Article 725).*

Motor control conductors as described above must be protected (usually with an in-line fuse) in accordance with column A of *Table 430.72(B)*, except:

1. When they extend no farther than the motor controller enclosure, they can be protected according to column B of *Table 430.72(B).*

2. When they extend farther than the motor controller enclosure, they can be protected according to column C of *Table 430.72(C).*

3. When control circuit conductors are taken from single-phase transformers that have only a 2-wire secondary and are considered to be protected by the protection on the primary side of the transformer. However, the primary protection ampacity shouldn't be more than the ampacity shown in *Table 430.72(B)* multiplied by the secondary-to-primary voltage ratio (secondary voltage divided by primary voltage).

4. When the opening of a control circuit would cause a hazardous situation (as would be the case with a fire pump, for example) and the control circuit can be tapped into the motor branch circuit with no further protection.

Control transformers must be protected according to *Article 450* or *Article 725* (see Chapters 21 and 38), except:

I. Control transformers that are an integral part of a motor controller and rated less than 50 VA can be protected by primary protective devices, impedance limiting means, or other means.

2. If the primary rating of the transformer is less than 2 amps, an overcurrent device rated at no more than 500% of the primary current can be used in the primary circuit.

3. By other approved means.

4. When the opening of a control circuit would cause a hazardous situation (as with a fire pump, for example), protection can be omitted.

When damage to a control circuit would create a hazard, the control circuit must be protected (by a raceway or other suitable means) outside of the control enclosure.

When one side of a motor control circuit is grounded, the circuit must be arranged so that an accidental ground won't start the motor.

Motor control circuits must be arranged so that they will be shut off from the current supply when the disconnecting means is in the open position.

Motor Controllers

Suitable controllers are required for all motors.

The branch-circuit protective device can be used as a controller for motors of ⅛ horsepower or less that are normally left running and can't be damaged by overload or failure to start.

Portable motors of ⅓ horsepower or less may use a plug-and-cord connection as a controller.

Controllers must have horsepower ratings no lower than the horsepower rating of the motor they control, except:

1. Stationary motors rated 2 horsepower or less and 300 volts or less can use a general-use switch that has an ampere rating at least twice that of the motor it serves. General-use ac snap switches can be used on ac circuits to control a motor rated 2 horsepower or less and 300 volts or less and having an ampere rating of no more than 80% of the switch rating.

2. A branch-circuit inverse-time circuit breaker that is rated in amperes only (no horsepower rating) can be used.

If a controller is also functioning as a disconnecting means, it does not have to open all conductors to the motor.

If power to a motor is supplied by a phase converter, the power must be controlled in such a way that, in the event of a power failure, power to the motor is cut off and can't be reconnected until the phase converter is restarted.

Each motor must have its own controller, except if:

A group of motors (600 volts or less) uses a single controller that is rated at no less than the sum of all motors connected to the controller. This applies only in the following cases:

1. When a group of motors drives several parts of a single machine.

2. When a group of motors is protected by one overcurrent device, as specified elsewhere.

3. When the group of motors is located in one room, and within sight of the controller.

A controller must be capable of stopping and starting the motor, and of interrupting its lock-rotor current.

The disconnecting means must be located within sight of the controller location and within sight of the motor, except in the following situations:

1. If the circuit is *over* 600 volts, the controller disconnecting means can be out of sight from the controller, as long as the controller has a warning label that states the location of the disconnecting means is locked in the open position.

2. One disconnecting means can be located next to a group of coordinated controllers on a multi-motor continuous process machine.

The disconnecting means for motors 600 volts or less must be rated at least 115% of the full-load current of the motor being served.

Controllers that operate motors over 600 volts must have the control circuit voltage marked on the controller.

Fault-current protection must be provided for each motor operating at over 600 volts (see *Section 430.125(C)*).

All exposed live parts must be protected (see *Sections 430.132* and *430.133*, if necessary).

Grounding

The frames of portable motors that operate at more than 150 volts must be grounded or guarded.

The frames of stationary motors must be grounded (or permanently and effectively isolated from ground) in the following circumstances:

When supplied by metal enclosed wiring.

In wet locations, when they are not isolated or guarded.

In hazardous locations.

If any terminal of the motor is over 150 volts to ground.

All controller enclosures must be grounded, except when attached to portable ungrounded equipment.

Controller-mounted devices must be grounded.

Chapter 19

HVAC Equipment

Air-Conditioning and Refrigeration Equipment (*Article 440*)

In applying other sections of the National Electrical Code, an air-conditioning or refrigeration system is to be considered a single machine.

Disconnecting Means

A disconnect for a hermetic refrigerant motor-compressor must be sized at 115% of the nameplate current or branch-circuit current, whichever is higher.

Disconnect switches rated over 100 horsepower must be marked "DO NOT OPERATE UNDER LOAD."

A cord-and-plug connection can be considered as a disconnecting means for equipment such as room air conditioners, household refrigerators and freezers, water coolers, etc.

The disconnecting means must be located within sight of the equipment it serves. It may be located on or in the equipment.

Branch Circuits

Branch-circuit ratings are marked on the nameplates of air-conditioning and refrigeration units (based on *Section 440.22*).

Branch-circuit conductors to a single motor-compressor must have an ampacity of at least 125% of the unit's rated current or the branch circuit's rated current, whichever is greater.

Conductors to more than one motor-compressor must be sized as follows:

> The rated load or branch-circuit rating (whichever is larger) of the largest motor-compressor must be added to the full-load currents of all other motors or motor-compressors, plus 25% of the highest-rated motor or motor-compressor in the group. Room air conditioners are excepted. In instances where control circuits are interlocked so that a second or group of motor-compressors can't be started when the largest is operating, the conductor size shall be determined from the rating(s) of the highest unit or units that can operate at any time.

When combination motor-compressor and lighting loads are connected to the same circuit, the conductor size must be based on the lighting load, plus the ampacity required for the motor-compressor(s). In instances where control circuits are interlocked so that motor-compressors and lighting loads can't be operated at the same time, the conductor size shall be determined from the rating of either the motor-compressor or the lighting load, whichever is higher.

Controllers and Overload Protection
Controllers and overload protection are virtually always built into HVAC units. In the rare case that they are not, the standards in *Sections 440.41, 440.51, 440.52, and 440.53* must be used. However, it is highly recommended that an experienced equipment designer make these determinations.

Room Air Conditioners
All room air conditioners must be grounded.

The circuits supplying room air conditioners should be sized to at least 125% of the circuit size shown on the unit's nameplate if no other equipment is supplied by the circuit. If other loads are also connected to the circuit, the circuit must be at least 200% of the unit's rating. An exception to the 200% rule (the air conditioner may be rated no more than 50% of the branch circuit) is made if the circuitry is interlocked to prevent the air conditioner from operating at the same time as other loads. In such cases, the air conditioner may use no more than 80% of the circuit's rated current (this is the same as rating the circuit to 125% of the unit's full load).

A plug-and-receptacle combination is allowed as the required disconnecting means for units of 250 volts or less, if the unit's controls are located within 6 feet from the floor or a manual switch is located within sight of the unit in an accessible location.

Cords may not be longer than 6 feet for 208- or 240-volt units, and 10 feet for 120-volt units.

Chapter 20

Generators

Generators (*Article 445*)

Locations
Generators must be of a suitable type for the areas in which they are installed.

Overcurrent Protection
Constant-voltage generators (which include virtually all in use) must have overcurrent protection provided by inherent design, circuit breakers, or other means. Alternating-current generator exciters are excepted.

Two-wire dc generators may have only one overcurrent device if it is actuated by the entire current generated, the current in the shunt field excepted. The shunt field is not to be opened.

Generators that put out 65 volts or less and are driven by an electric motor are considered protected if the motor driving them will trip its overcurrent protection device when the generator reaches 150% of its full-load rated current.

Installation
The ampacity of conductors from the generator terminals to the first overcurrent protection device must be at least 115% of the generator's nameplate rated current. This applies only to phase conductors; neutral conductors can be sized for only the load they will carry (see *Section 220.22*).

Live parts operating at more than 50 volts must be protected. Guards are to be provided where necessary.

Bushings must be used where wires pass through enclosure walls.

Chapter 21

Transformers

Transformers (*Article 450*)

In this volume, the term *transformer* refers to a single transformer, whether it be single-phase or polyphase.

Overcurrent Protection

Transformers operating at over 600 volts must have protective devices for both the primary and secondary of the transformer. Their sizes must be in accordance with *Table 450.3(A)(1)*. If the specified fuse or circuit breaker rating does not correspond to a standard rating, the next larger size can be used.

Transformers operating at over 600 volts that are overseen only by qualified persons can be protected in accordance with the lower-right sections of *Table 450.3(A)*. If the specified fuse or circuit breaker rating does not correspond to a standard rating, the next larger size can be used.

Transformers rated 600 volts or less can be protected by an over-current protection device on the primary side only, which must be rated at least 125% of the transformer's rated primary current. If the specified fuse or circuit breaker rating for transformers with a rated primary current of 9 amperes or more does not correspond to a standard rating, the next larger size can be used. For transformers with rated primary currents of less than 9 amperes, the overcurrent device can be rated up to 167% of the primary rated current (Figure 21.1).

Transformers operating at less than 600 volts are allowed to have overcurrent protection in the secondary only, which must be sized at 125% of the rated secondary current, *if* the feeder overcurrent device is rated at no more than 250% of the transformer's rated primary current. Transformers with thermal overload devices in the primary side don't require additional protection in the primary side unless the feeder overcurrent device is more than six times the primary's rated current (for transformers with 6% impedance or less), or four times primary current (for transformers with between 6% and 10% impedance). If the specified fuse or circuit breaker rating for transformers with a rated primary current of

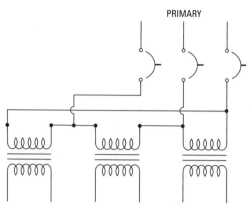

PRIMARY

Figure 21.1 Overcurrent protection may be at the transformer or in the circuit supplying the transformer.

9 amperes or more does not correspond to a standard rating, the next larger size can be used. For transformers with rated primary currents of less than 9 amperes, the overcurrent device can be rated up to 167% of the primary rated current.

Potential transformers must have primary fuses.

Autotransformers rated 600 volts or less must be protected by an overcurrent protection device in each ungrounded input conductor, which must be rated at least 125% of the rated input current. If the specified fuse or circuit breaker rating for transformers with a rated input current of 9 amperes or more does not correspond to a standard rating, the next larger size can be used. For transformers with rated input currents of less than 9 amperes, the overcurrent device can be rated up to 167% of the rated input current.

Transformers that are dedicated to fire pumps don't require secondary overcurrent protection. The primary overcurrent protection must be sufficient to carry the total of all locked rotor and associated currents fed by it for an indefinite period of time. This is necessary because fire pumps operate only under extreme emergency conditions, and in such cases the circuit tripping an overcurrent device could lead to an unrestrained fire. Since the risks due to a fire pump not functioning are greater than the risks from locked rotor currents, it is better to allow for the locked rotor currents, and guarantee that the fire pumps will function without interruption.

Installation

Transformers must be installed in places that have enough ventilation to avoid excessive heat buildup.

All exposed non–current-carrying parts of transformers must be grounded.

Transformers must be located in accessible locations, except as follows:

> Dry-type transformers operating at less than 600 volts that are located in the open on walls, columns, and structures don't have to be in accessible locations. (See definition of "Accessible" in glossary.) Dry-type transformers operating at less than 600 volts and less than 50 VA are allowed in fire-resistant hollow spaces of buildings, as long as they have enough ventilation to avoid excessive heating.

Indoor dry-type transformers rated 112.5 kVA or less must be separated by at least 12 inches from combustible materials. Fire-resistant, heat-resistant barriers can be substituted for this requirement. Also, such transformers operating at 600 volts or less that are completely enclosed are exempt from this requirement.

Indoor dry-type transformers rated over 112.5 kVA must be installed in rooms made of fire-resistant materials. Such transformers with 80° C or higher rating can be separated from combustible materials by fire-resistant, heat-resistant barriers, or may be separated from combustible materials by at least 6 feet horizontally or 12 feet vertically. Also, transformers with 80° C or higher ratings that are completely enclosed (except for ventilated openings) are exempt from this requirement.

Dry-type transformers installed outdoors must be installed in rain-tight enclosures and must not be located within 12 inches of combustible parts of buildings. Transformers with 80° C or higher ratings that are completely enclosed (except for ventilated openings) are exempt from this requirement.

Materials can't be stored in transformer vaults.

Chapter 22

Capacitors, Resistors, and Batteries

Capacitors (*Article 460*)

Conductors
The ampacity of conductors for capacitor circuits must be at least 135% of the rated current of the capacitor being served. Conductors connecting capacitors to motors or motor circuits must be rated no less than one-third the full-load motor current or 135% of the capacitor's rated current, whichever is greater.

Overcurrent protection must be provided in each ungrounded conductor in each capacitor installation. Overload devices are *not* required for conductors on the load side of motor overload protection devices. The rating of such devices should be as low as is practical.

A disconnecting means must be provided in each ungrounded conductor in each capacitor installation. A disconnecting means is *not* required for conductors on the load side of motor overload protective devices. The rating of such a device may not be lower than 135% of the capacitor's rated current. The disconnecting means must open all ungrounded conductors, and may disconnect the capacitor from the line as a regular operating procedure.

Capacitor cases must be grounded.

Resistors and Reactors (*Article 470*)
Only individual resistors and reactors are covered by this article, not components that are parts of other pieces of equipment.

Installation
Resistors and reactors must not be located where they can be subjected to physical damage.

Resistors and reactors must either be placed 12 inches or more from combustible materials, or have a thermal barrier installed between them and any combustible materials.

Resistors or reactors operating at over 600 volts must be elevated or enclosed to avoid accidental contact with energized parts.

Resistors and reactors operating at over 600 volts must be placed at least 12 inches from combustible materials.

Reactor or resistor cases must be grounded.

Storage Batteries (*Article 480*)

Storage battery installations must be carefully designed. Do *not* think that storage batteries are relatively harmless because of your familiarity with them as automobile batteries. Remember, most design requirements are not shown here. Do *not* install storage batteries without a proper design layout: they can be explosive, and very dangerous.

Installation

Wiring to and from storage batteries must conform with all applicable parts of the NEC.

Racks required for mounting storage batteries must be made of treated metal that will resist corrosion from the electrolyte in the batteries and with nonconductive parts directly supporting the batteries, or made of fiberglass or the like.

Trays fitting into battery racks must be made of material that will resist the deteriorating action of the electrolyte.

Batteries must be installed in locations where they are well ventilated, and where live parts will be guarded.

Vented cells must be provided with flame arrestors.

Sealed cells must be equipped with pressure release vents.

Chapter 23

Hazardous Locations

Hazardous Locations (*Articles 500, 501, 502, 503, 504*)

All electrical installations in hazardous locations are inherently dangerous. Do *not* perform installations without carefully engineered layouts. If you don't have first-rate instructions, *do not* install the wiring! The installation requirements in this chapter are given to assist in the installation process, *not* to serve as a substitute for an engineered layout. This work can be dangerous—don't take chances.

General

Locations are classified as hazardous depending on the nature of the chemicals, dust, or fibers that may be present; and also upon their concentrations in the various environments.

In determining classifications, each room or area is considered separately.

Intrinsically safe circuits can be installed in any hazardous location, but must be kept isolated from all other wiring systems that are not intrinsically safe.

All equipment installed in hazardous areas (also called *classified* areas) must be approved for the *specific area* in which it is installed, not just approved for hazardous locations in general.

The wiring requirements for one type of hazardous location can't be substituted for the requirements for another type of location. They are not interchangeable.

Locknut-bushing or double-locknut connections are not considered adequate bonding methods for hazardous locations. Other methods must be used.

With the use of ingenuity in laying out wiring for hazardous locations, a great deal of the wiring can be located outside of the hazardous areas, entering into the hazardous areas only where necessary. This can avoid a great deal of the cost and difficulty of these installations.

Class I Locations

Class I locations are areas where flammable gases or vapors are present in amounts great enough to produce explosive or ignitable mixtures.

Class I, division 1 locations are areas where flammable concentrations of gases or vapors may be present under normal operating conditions, or in which such gases are frequently present because of maintenance or leakage, or where a breakdown might cause such vapors or gases to be present.

Class I, division 2 locations are areas in which flammable liquids, vapors, or gases are handled or processed, but in which the liquids, vapors, or gases are normally held in closed containers from which they will escape only in abnormal circumstances.

Groups in class I are determined by the ignition temperatures for the types of gases or vapors in them.

Wiring methods in class I, division 1 areas may be any of the following:

> Threaded rigid metal conduit.
>
> Threaded intermediate metal conduit.
>
> Type MI cable, using fittings that are suitable for the location.

All boxes, fittings, and joints must be threaded for conduit and cable connections, and must be explosion-proof.

Threaded joints must be made up with at least 5 threads fully engaged.

MI cable must be carefully installed so that no tensile (pulling) force is placed on the cable connectors.

Flexible connections can be used only where necessary, and must be made with materials approved for class I locations.

Seals must be provided within 18 inches of any enclosure housing equipment that can produce sparks, such as switches, relays, circuit breakers, etc. (Figures 23.1, 23.2).

Seals must also be provided where conduits enter or leave a class I, division 1 area. The seals may be on either side of the dividing wall (Figure 23.3).

Wiring methods in class I, division 2 areas may be any of the following:

> Threaded rigid metal conduit.
>
> Threaded intermediate metal conduit.
>
> Type MI, MV, MC, TC, or SNM cables, using fittings that are suitable for the location. These cables must be carefully installed so that no tensile (pulling) force is placed on the cable connectors.
>
> Enclosed gasketed busways.

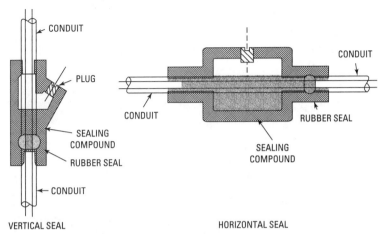

VERTICAL SEAL   HORIZONTAL SEAL

Figure 23.1   Horizontal and vertical seals.

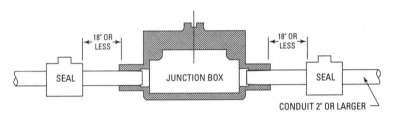

Figure 23.2   Junction boxes with splices or taps shall have seals.

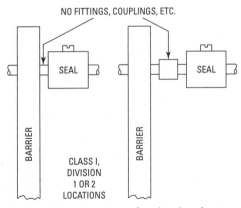

Figure 23.3   Location of seals when leaving class I locations.

Enclosed gasketed wireways.

PLTC cable, installed under the provisions of *Article 725*.

MI, MV, MC, TC, SNM, and PLTC cables may be installed in cable trays.

All boxes, fittings, and joints must be threaded for conduit and cable connections, and must be explosion-proof.

Threaded joints must be made up with at least 5 threads fully engaged.

Flexible connections can be used only where necessary, and must be made with materials approved for class I locations.

Seals must be provided within 18 inches of any enclosure housing equipment that can produce sparks, such as switches, relays, circuit breakers, etc.

Seals must also be provided where conduits enter or leave classified areas. The seals may be on either side of the dividing wall.

Class II Locations

Class II locations are areas in which combustible dust may be present.

Class II, division 1 locations are areas in which the concentration of flammable dust under normal conditions is sufficient to produce an explosive or ignitable mixture.

Class II, division 2 locations are areas in which flammable or ignitable dust is present, but normally not in quantities sufficient to produce a flammable or explosive mixture.

Wiring methods in class II, division 1 areas may be any of the following:

Threaded rigid metal conduit.

Threaded intermediate metal conduit.

Type MI cable, using fittings that are suitable for the location.

All boxes, fittings, and joints must be threaded for conduit and cable connections, and must be approved for class II locations (Figure 23.4).

MI cable must be carefully installed so that no tensile (pulling) force is placed on the cable connectors.

Flexible connections can be made by any of the following means:

Liquid-tight flexible metal conduit, with approved fittings.

Liquid-tight flexible nonmetallic conduit, with approved fittings.

Extra-hard-usage cord, with bushed fittings and dust seals.

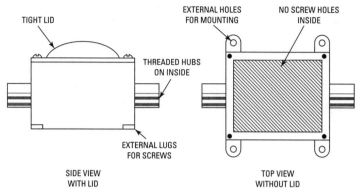

TIGHT LID

EXTERNAL HOLES FOR MOUNTING

NO SCREW HOLES INSIDE

THREADED HUBS ON INSIDE

EXTERNAL LUGS FOR SCREWS

SIDE VIEW WITH LID

TOP VIEW WITHOUT LID

Figure 23.4 Type of box for class II locations.

When raceways extend between class II, division 1 locations and unclassified locations, sealing may be done in any of the following ways (Figure 23.5):

With raceway seals.

With a 10-foot horizontal run of raceway.

With a 5-foot vertical raceway.

Wiring methods in class II, division 2 locations may be any of the following:

Threaded rigid metal conduit.

Threaded intermediate metal conduit.

Type MI, MC, or SNM cables, using fittings that are suitable for the location. These cables must be carefully installed so that no tensile (pulling) force is placed on the cable connectors.

Enclosed gasketed busways.

Enclosed gasketed wireways.

MC, TC, and PLTC cables may be installed in cable trays.

When raceways extend between class II, division 2 locations and unclassified locations, sealing may be done in any of the following ways:

With raceway seals.

With a 10-foot horizontal run of raceway.

With a 5-foot vertical raceway.

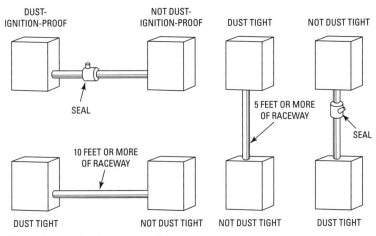

Figure 23.5 Sealing in class II locations.

Flexible connections can be made by any of the following means:

Liquid-tight flexible metal conduit, with approved fittings.

Liquid-tight flexible nonmetallic conduit, with approved fittings.

Extra-hard-usage cord, with bushed fittings and dust seals.

Class III Locations

Class III locations are termed hazardous areas because of the presence of easily ignitable fibers or flyings, but where these fibers or flyings are not likely to be suspended in the air in quantities sufficient to cause an ignitable mixture.

Class III, division 1 locations are areas where easily ignitable fibers or materials producing easily ignitable flyings are handled, manufactured, or used.

Class III, division 2 locations are areas where ignitable fibers are stored or handled.

Wiring methods in class III, division 1 or 2 locations may be any of the following:

Threaded rigid metal conduit.

Threaded intermediate metal conduit.

Type MI, MC, or SNM cables, using fittings that are suitable for the location. These cables must be carefully installed so that no tensile (pulling) force is placed on the cable connectors.

Dust-tight wireways.

Flexible connections can be made by any of the following means:

Liquid-tight flexible metal conduit, with approved fittings.
Liquid-tight flexible nonmetallic conduit, with approved fittings.
Extra-hard-usage cord, with bushed fittings and dust seals.

Intrinsically Safe Systems

Intrinsically safe systems must be designed for the specific installation, complete with control drawings. This is required by the NEC, and no intrinsically safe system installation may be attempted without such drawings.

Commercial Garages (*Article 511*)

In commercial garages, the entire area of the garage, from the floor up to a height of 18 inches, must be considered a class I, division 2 location (Figure 23.6).

Any pit or depression in the floor shall be considered a class I, division 1 location from the floor level down (Figure 23.7).

Adjacent areas such as storerooms, switchboard rooms, etc. are not considered to be classified areas if they have ventilating systems that provide four or more changes per hour, or if they are well separated from the garage area by walls or partitions.

Areas around fuel pumps are covered by *Article 514*.

Wiring methods in areas above class I locations may be any of the following:

Rigid metal conduit.
Intermediate metal conduit.
Rigid nonmetallic conduit.
Electrical metallic tubing.
MI, TC, SNM, or MC cables.

Plug receptacles above class I locations must be approved for the purpose.

Electrical equipment that could cause sparks located above class I locations must be totally enclosed, or located at least 12 feet from the floor.

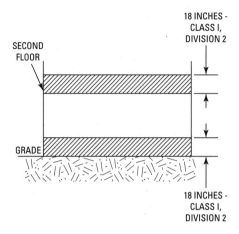

Figure 23.6 Hazardous areas of commercial garages above grade level.

All receptacles installed where hand tools, diagnostic equipment, or portable lighting devices are to be used must have ground-fault protection.

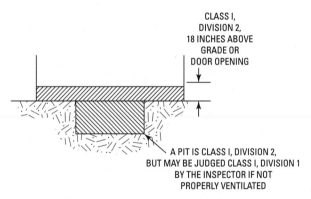

Figure 23.7 Hazardous areas and classification of pits or depressions in garage floors.

Airplane Hangars (*Article 513*)

In airplane hangars, the entire area from the floor up to a height of 18 inches must be considered a class I, division 2 location (Figure 23.8).

Any pit or depression in the floor shall be considered a class I, division 1 location from the floor level down.

Adjacent areas such as storerooms, switchboard rooms, etc. are not considered classified areas if they are well separated from the garage area by walls or partitions.

Wiring methods in areas above class I locations may be any of the following:

Rigid metal conduit.

Intermediate metal conduit.

Electrical metallic tubing.

MI, TC, SNM, or MC cables.

Electrical equipment that could cause sparks located above class I locations must be totally enclosed, or located at least 10 feet from the floor.

Aircraft electrical systems must be deenergized when an aircraft is stored in a hangar and, whenever possible, when the aircraft is being serviced.

Aircraft batteries can't be charged when they are installed in an aircraft located partially or fully inside a hangar.

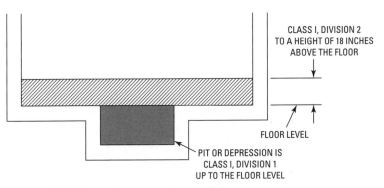

Figure 23.8   Hazardous areas in aircraft hangars.

Chapter 24

Service Stations

Gasoline Dispensing and Service Stations
(*Article 514*)

Gasoline dispensing and service stations are hazardous locations, and as stated in Chapter 23, there are dangers involved with wiring in these locations. No installations should be done in these areas without carefully engineered drawings. The requirements presented here are for informational purposes and don't replace a properly engineered layout.

Installation

Tables 514.3(B)(1) and *514.3(B)(2)* specify which locations in service stations are considered classified. Wiring in these areas must be in accordance with the applicable class and division.

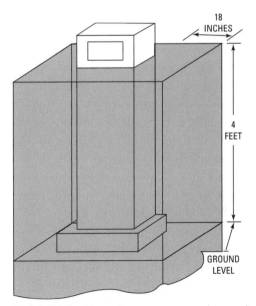

Figure 24.1   Hazardous areas around a gasoline dispenser.

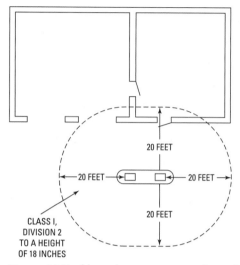

Figure 24.2   Hazardous areas around gasoline service stations.

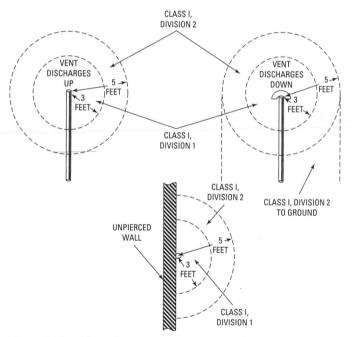

Figure 24.3   Hazardous areas around vent pipes.

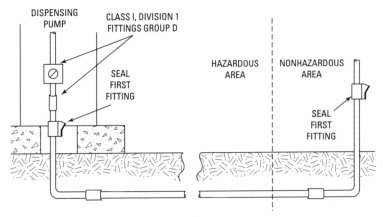

Figure 24.4   Seals required in gasoline service stations.

Underground wiring must be in either threaded rigid metal conduit or threaded intermediate metal conduit. Any portion under a class I, division 2 area must be considered a class I, division 1 location, and shall be considered as such up to the point at which the raceway emerges from the ground or floor. Properly installed MI cable is also permitted. Also, rigid nonmetallic conduit can be used if at least 2 feet below grade, and if rigid metal conduit is used for the last 2 feet of the run prior to its emergence from the ground or floor (Figures 24.1, 24.2, 24.3, 24.4).

Chapter 25

Bulk Storage Plants

Bulk Storage Plants (*Article 515*)

Bulk storage plants are hazardous locations, and as stated in Chapter 23, there are dangers involved with wiring in these locations. No installations should be done in these areas without carefully engineered drawings. The requirements shown here are for informational purposes and don't replace a properly engineered layout.

Installation

Table 515.3 specifies which locations in bulk storage plants are considered classified. Wiring in these areas must be in accordance with the applicable class and division.

Underground wiring must be in either threaded rigid metal conduit or threaded intermediate metal conduit. Any portion under a class I, division 2 area must be considered a class I, division 1 location, and shall be considered as such up to the point at which the raceway emerges from the ground or floor. Properly installed MI cable is also permitted. Also, rigid nonmetallic conduit can be used if at least 2 feet below grade, and if rigid metal conduit is used for the last 2 feet of the run prior to its emergence from the ground or floor.

All wiring above class I locations must be done by one of the following methods:

Rigid metal conduit.

Intermediate metal conduit.

Electrical metallic tubing.

Schedule 80 PVC conduit.

Type MI, MC, TC, or SNM cables.

Equipment above class I areas must be totally enclosed (Figures 25.1, 25.2, 25.3, 25.4, 25.5, 25.6).

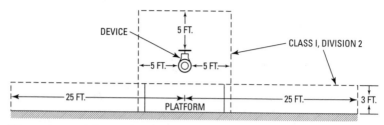

Figure 25.1 Adequately ventilated indoor areas.

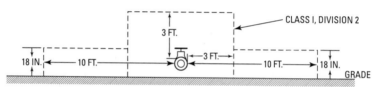

Figure 25.2 Outdoor areas.

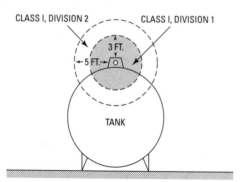

Figure 25.3 Classification around vents.

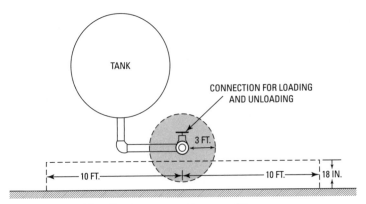

Figure 25.4 Classification around bottom filler.

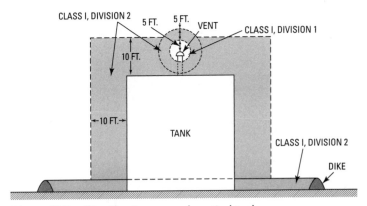

Figure 25.5 Classification around vertical tank.

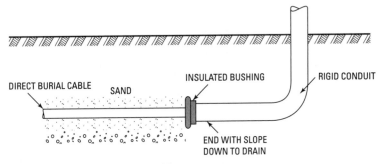

Figure 25.6 Direct-burial cable.

Chapter 26

Spray Areas

Spray Application, Dipping, and Coating Processes (*Article 516*)

Spray areas are hazardous locations, and as stated in Chapter 23, there are dangers involved with wiring in these locations. No installations should be done in these areas without carefully engineered drawings. The requirements shown here are for informational purposes and don't replace a properly engineered layout.

Installation

Figure 26.1 specifies which locations in spraying areas are considered classified. Wiring in these areas must be in accordance with the applicable class and division.

Adjacent areas such as storerooms, switchboard rooms, etc. are not considered to be classified areas if they are well separated from the spray area by tight walls or partitions without communicating openings.

All wiring above class I locations must be done by one of the following methods:

Rigid metal conduit.

Intermediate metal conduit.

Electrical metallic tubing.

Rigid nonmetallic conduit.

Type MI, MC, TC, or SNM cables.

Equipment above class I areas must be totally enclosed (Figure 26.1).

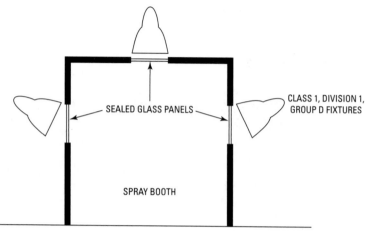

SEALED GLASS PANELS

CLASS 1, DIVISION 1,
GROUP D FIXTURES

SPRAY BOOTH

Figure 26.1   Approved lighting for spray booths.

Chapter 27

Health Care Facilities

Health Care Facilities (*Article 517*)

It is very important to wire all health care facilities exactly according to carefully prepared and engineered drawings, not only for safety's sake, but also because of the very exacting requirements of the Health and Rehabilitation Services administration (HRS). HRS has its own set of requirements and performs its own inspections. *Never* begin an installation in a health care facility without engineered drawings; the HRS is *very* serious about its requirements, and the author has seen instances of electrical contractors having to bring their work up to HRS's standards at costs of up to $7 million. The companies were *not* reimbursed for this work.

General Areas

These areas include general health care areas, including such areas as a doctor's examining room in a multifunction building. These areas don't include business offices, waiting rooms, or patient rooms in nursing homes.

Except where specifically prohibited (which is common), any applicable types of wiring can be used in these areas.

In patient care areas, all exposed metal surfaces and grounding terminals of receptacles must be grounded with an insulated copper grounding conductor. This conductor must be sized according to *Table 250.122.*

Wiring in patient care areas must use one or more of the following methods:

Rigid metal conduit.

Intermediate metal conduit.

Electrical metallic tubing.

MI, MC, or AC cable.

Equipment grounding busses in panelboards for essential and normal branch circuits must be bonded with at least a No. 10 AWG insulated copper conductor.

Ground-fault protection is required in patient care areas, for both branch circuits and feeders.

Bed locations must be provided with at least two branch circuits—one from the emergency system and one from the normal system.

The emergency circuit must supply at least one receptacle for that bed location only. The receptacle(s) must be identified, as also must be the panelboard and circuit number.

All circuits on the normal power system must be from a single panelboard.

All circuits on the emergency power system must be from a single panelboard.

All patient bed locations must have at least six hospital-grade receptacles. All these receptacles must be bonded to an equipment-grounding point with a bonding jumper of at least No. 10 AWG insulated copper.

All panelboards must be grounded by one of the following means:

Where a locknut-bushing connection is used, grounding bushings and continuous copper bonding jumpers must be used. (Size jumper per *Section 250.66*.)

By connecting feeder raceways or cables to threaded hubs on the terminating enclosures.

Other approved devices such as bonding locknuts, etc.

Special design requirements for critical care areas, essential electrical systems, anesthetizing locations, X-ray locations, signaling systems, and isolated power systems are not given here.

Patient Care Areas

In patient care areas of health care facilities, only metal raceways or armored cables listed as a ground return path containing an insulated copper equipment-grounding conductor may be used for branch-circuit wiring supplying equipment, luminaires, and receptacles.

Chapter 28

Places of Assembly, Theaters, Motion Picture and Television Studios

Places of Assembly (*Article 518*)

Places of assembly are buildings, or parts of buildings, that are intended to be places of assembly for 100 or more persons.

Temporary wiring in these areas must conform to *Article 527* (see Chapter 7), except that cords can be laid on floors if they are protected from contact by the public.

Acceptable wiring methods in places of assembly are the following:

Rigid metal conduit.

Intermediate metal conduit.

Electrical metallic tubing.

Rigid nonmetallic conduit and other nonmetallic raceways, if encased in 2 inches or more of concrete.

MI or MC cables.

The numbers of conductors allowed in raceways is the same for places of assembly as for all standard types of wiring, with one exception: The limit of 30 current-carrying conductors in an auxiliary gutter or wireway does not apply.

Portable switchboards and power distribution systems can be supplied only from listed outlets that are sufficient in voltage and current. These outlets must have listed circuit breakers or fuses in accessible locations, and must have provisions for the connection of grounding conductors. The neutral of any feeder supplying 3-phase 4-wire dimming systems must be rated as a current-carrying conductor.

Theaters (*Article 520*)

Temporary wiring in these areas must conform to *Article 527* (see Chapter 7), except that cords can be laid on floors if they are protected from contact by the public.

Acceptable wiring methods in theaters are the following:

Rigid metal conduit.

Intermediate metal conduit.

Electrical metallic tubing.

Rigid nonmetallic conduit and other nonmetallic raceways, if encased in 2 inches or more of concrete.

MI or MC cables.

The number of conductors allowed in raceways is the same for theaters as for all standard types of wiring, with one exception: The limit of 30 current-carrying conductors in an auxiliary gutter or wireway does not apply.

All live parts (including parts of lighting switchboards) must be protected from accidental contact by persons or objects.

Circuits for floodlights, border lights, and proscenium side-light can't have an ampacity of greater than 20 amps, unless heavy-duty lampholders are used. Conductors to such lighting must have insulation capable of withstanding the heat encountered. The conductor insulation may not be rated less than 125°C.

Requirements for specific types of stage equipment are not provided here.

Only types of cable that are listed for extra-hard usage can be used to supply border lights. For ampacities of these cords, refer to *Table 520.44.*

Motion Picture and Television Studios (*Article 530*)

Permanent wiring in these areas must be either MC cable, MI cable, or approved raceways.

Portable wiring in these areas must be done with approved cord or cables. Splices are allowed if done by approved methods, and if the overcurrent protection to the cable in question is no more than 20 amperes.

Switches used for stage set lighting must be externally operable.

Receptacles for dc plugging boxes must be rated at least 30 amperes.

All live parts must be protected from accidental contact by persons or objects.

Overcurrent protection levels for various types of equipment in motion picture or TV studios are as follows:

Stage cables may be protected at up to 400% of their rated ampacity.

Feeders that go only from substations to stages, and operate only for periods of 20 minutes or less, can be protected for up to 400% of their rated ampacity.

Feeders to studios can be sized based on the derating factors of *Table 530.19(A)*.

All non–current-carrying parts of all equipment must be grounded.

Enclosed and gasketed lighting fixtures must be used in film storage areas. No other equipment is allowed in these areas.

Switches controlling the lights in film storage areas must be located outside the areas, must disconnect all ungrounded conductors (no standard 3-way switches), and must have a pilot light that indicates when the lights in the storage area are on.

Requirements for projectors are omitted here.

Chapter 29

Signs

Installation

Each outline lighting installation or sign (except portable signs) must be controlled with an externally operated switch or circuit breaker, which must open all ungrounded conductors.

The disconnecting means mentioned above must be within sight of the sign it protects. If the sign has an electronic or electromechanical controller not located in the sign, the disconnecting means can be installed within sight of the controller, rather than within sight of the sign.

Control devices that control transformers must be rated for an inductive load or amperage rating at least twice that of the transformer.

Wiring to signs must terminate in junction boxes.

Electric signs must be listed.

All metal parts of signs (except isolated parts) must be grounded.

Circuits to lamps, ballasts, and transformers can't be rated at more than 20 amperes.

Circuits that feed electric-discharge lighting transformers can't be rated at more than 30 amperes.

A commercial building accessible to the public must have at least one sign outlet on its own 20-amp circuit.

Flashers, cutouts, etc. must be installed in metal boxes that can be accessed after the sign is installed.

A weatherproof receptacle and attachment plug must be provided for each portable or mobile sign.

Cord used for connecting attachment plugs to portable signs must be a hard-usage or junior hard-usage type, per *Table 400.4*, and must have an equipment-grounding conductor.

Signs must be installed so that their lowest part is at least 16 feet above areas that are open to vehicular traffic. They can be mounted lower if suitably protected.

Outdoor portable signs must be ground-fault protected. The protective device can either be in the sign or in the power system that supplies the sign.

Any normal wiring method can be used for supplying signs operating at 600 volts or less.

Requirements for the manufacturing of electric signs are not included here.

Chapter 30

Manufactured Wiring Systems

Manufactured Wiring Systems (*Article 604*)

Installation

Manufactured wiring systems can be installed in dry, accessible locations and, when listed for this use, in plenums and spaces for environmental air.

It is allowed to extend one end of a cable into a hollow wall when necessary to reach a switch or outlet.

Manufactured wiring systems are made from type AC or type MC cable, and are subject to any restrictions placed on these cable types.

All unused outlets must be capped.

The entire system must be properly grounded.

Office Furnishings (*Article 605*)

Installation

Wiring in office furnishings (usually partitions) must be of a type for which the furnishings are listed.

All wiring and connections must be done in the wiring channels of the office furnishings that are provided for this use.

Electrical connections between partitions must be made with types of wiring that are listed for these partitions, or with flexible cords if all of the following conditions are met:

Extra-hard-usage cord must be used.

The partitions must be mechanically secured to each other and continuous.

The cord must be only as long as necessary, but never longer than 2 feet.

The cord must terminate in an attachment plug-and-cord connector with strain relief.

Any lighting used with such partitions must be listed for this purpose.

Where lighting such as that mentioned above uses cord-and-plug connections, the conductors in the cord must be at least No. 18 AWG copper, and the cords can't be longer than 9 feet.

Fixed partitions must be permanently connected to the building wiring system.

Free-standing partitions can be connected to the building electrical system, but are not required to be so connected.

Free-standing partitions that are mechanically continuous and no more than 30 feet long can be connected to the building electrical system with a flexible cord. Such a cord must be no more than 2 feet long, and must be an extra-hard-usage cord with conductors no smaller than No. 12. The cord must have an insulated grounding conductor.

A receptacle supplying power to a partition as mentioned above must be on a separate circuit with no other loads connected to it. It must be within 12 inches of the partition(s) it serves.

Groups of partitions may not have more than 13 receptacles in them.

Partitions or groups of partitions are not allowed to contain multi-wire circuits, such as a 3-phase, 4-wire network.

Chapter 31

Mobile Homes and RV Parks

Mobile Homes (*Article 550*)

Requirements for the construction of mobile homes are omitted here.

Service Equipment

Service equipment for mobile homes must be located outside (not in or on) the mobile home. It must be located within sight and not more than 30 feet away, except if all of the following requirements are met (Figures 31.1, 31.2, 31.3, 31.4):

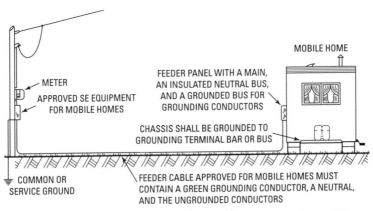

Figure 31.1 Pole-mounted service-entrance equipment for mobile home use with the feeder cable above ground.

1. A disconnecting means suitable for use as service equipment is located no more than 30 feet from the mobile home it serves, and also within sight.

2. A grounding electrode (see *Article 250* and Chapter 5) is present or installed at the disconnecting means.

3. A grounding electrode conductor connects the equipment-grounding terminal of the disconnecting means to the grounding electrode conductor.

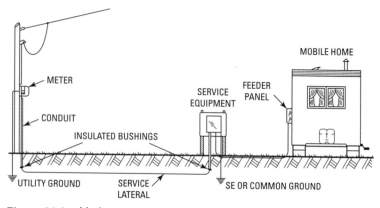

Figure 31.2 Underground service lateral for use with a mobile home.

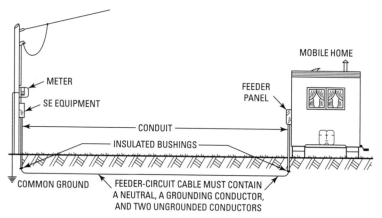

Figure 31.3 Pole-mounted service-entrance equipment for mobile home use with the feeder cable buried.

Service equipment for mobile homes can't be rated less than 100 amperes.

The outdoor mobile home disconnecting means must be mounted so that the bottom of the enclosure is at least 2 feet above grade and the operating handle is no more than 6.5 feet above grade.

All mobile home services must be grounded.

Mobile home lot feeder conductors must be suitable for the loads they must carry, and may never have ampacities less than 100 amperes.

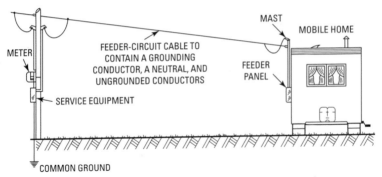

Figure 31.4   An overhead feeder cable installation to supply power to a mobile home.

A mobile home feeder between the service equipment and the disconnecting means does not need to include an equipment-grounding conductor if the neutral is grounded in accordance with *Section 250.24(A)*.

Power Supply

When its calculated load does not exceed 50 amperes, the power supply to a mobile home must be a 50-ampere mobile home supply cord with an integral cap, or a permanently installed feeder. Mobile homes with gas- or oil-fired cooking and heating systems are allowed to use a 40-amp cord assembly.

The cord mentioned above must be a 4-conductor cord permanently connected to the distribution panel, or to a junction box that is connected to the distribution panel.

The attachment cap plug must be a 50-amp, 4-wire, 3-pole, grounding-type cap.

The power supply cord must be between 21 feet and 36.5 feet long. The distance between the plug-in location and the mobile home must be at least 20 feet.

The power supply cord must be marked as suitable for mobile homes and must state the amperage.

The feeder assembly (cord or other type) must enter the mobile home through the floor, exterior wall, or roof.

Where the cord passes through floors or walls, bushings or conduits must be used to protect the cord. Cords may be run through walls between the panel and the floor, but in these cases, they must be installed in conduits sized at least 1¼-inch (trade size).

In cases where the calculated load exceeds 50 amperes or where permanently installed feeders are used for other reasons, the mobile home can be supplied by one of the following means:

> An overhead mast weatherhead service (see *Article 230* and Chapter 4) with four insulated, color-coded conductors. One of these conductors must be an equipment-grounding (green) conductor.

> A metal raceway or rigid nonmetallic conduit from the disconnecting means in the mobile home to the underside of the mobile home.

Disconnecting Means

A disconnecting means can be combined with the branch-circuit protective devices as a single assembly—a main circuit breaker panel. The main breaker is the disconnecting means.

Where the above method is not used, a single circuit breaker or fused switch must be used as the disconnecting means.

The disconnecting means must be marked "MAIN."

The panel must have a grounding terminal bar, which must be of sufficient size for the connection of all grounding conductors.

In mobile homes, the neutral conductors must be insulated from all equipment-grounding conductors. The neutral bus must be insulated from the equipment enclosure and all other grounded parts.

The distribution panel must be installed in an accessible location, but not in a bathroom. Working space 30 inches deep and 30 inches wide must be provided in front of the distribution panel.

The disconnecting equipment must be mounted at least 24 inches above the mobile home's floor.

Branch Circuits

Branch circuits in mobile homes are subject to the following restrictions:

> They may not be rated higher than the ampacity of the branch-circuit conductors.

> They may not be rated more than 150% of a single appliance rated 13.3 amperes or higher, supplied by an individual branch circuit.

> They can't be more than the size specified on the nameplate of the air conditioner or other motor-operated appliance being served by the circuit.

Branch circuits on the underside of the mobile home must be installed by one of the following wiring methods:

Rigid metal conduit.

Intermediate metal conduit.

Electrical metallic tubing, only when the tubing closely follows the frame.

Rigid nonmetallic conduit, only when the tubing closely follows the frame.

Throughout any mobile home, the neutral conductor must be kept isolated from the equipment-grounding conductors.

Wiring to standard ranges and clothes dryers must be made with 4-conductor cord and with 3-pole, 4-wire grounding-type plugs; or with type AC or MC cables, or with conductors enclosed with flexible metal conduit. In any of these cases, the neutral and grounding conductors must be isolated one from another. Standard range cable (3-wire) with a concentric (uninsulated) neutral may *not* be used.

Receptacles

A 15-amp multiple receptacle (not single receptacle) is allowed to be installed on a 20-amp laundry circuit.

All receptacles in mobile homes must be of the standard grounding type.

Any receptacles installed within 6 feet of a sink or lavatory must be protected by a ground-fault circuit interrupter. Any receptacles in a bathtub or shower area must also be protected by ground-fault interrupters.

Receptacles are forbidden within 30 inches of a tub or shower.

If heat tape outlets are installed, they must be mounted within 2 feet of the cold water pipe for which they provide protection.

Grounding

All exposed non–current-carrying metal parts in mobile homes must be grounded to the equipment-grounding bus in the panelboard, *not* to the neutral bus.

A bonding conductor must be installed between the equipment-grounding conductor bus and the chassis of the mobile home.

Recreational Vehicle Parks (*Article 551*)

Requirements for the manufacture of recreational vehicles are not covered here.

Electric Service at Sites

Each recreational vehicle site that is furnished with electricity must have at least one 15- or 20-amp, 125-volt receptacle.

At least 75% of all recreational vehicle sites having electric supply in an RV park must be equipped with a 30-amp, 125-volt, 2-pole, 3-wire grounding-type receptacle.

Other receptacles may also be provided at these locations, including 50-amp, 3-pole, 4-wire, 125/250-volt grounding receptacles.

All 15- and 20-amp, 125-volt receptacles must have ground-fault circuit interrupter protection.

Distribution

The secondary power distribution system in recreational vehicle parks must be a single-phase, 120/240-volt, 3-wire system.

Feeders to each site must have adequate capacity for the load being served. In no case may the ampacity be less than 30 amps. Because of the long distances involved in RV parks, it is almost always necessary to use larger sizes of wire than the ones listed in the tables of *Article 310*. This is necessary to avoid excessive voltage drops, which shouldn't exceed 5% at the outlet.

Overcurrent protection must be provided.

All electrical equipment in RV parks must be grounded.

Distribution systems must be grounded at their transformers.

The neutral conductor can't be used as an equipment ground.

No connection between a neutral and a grounding electrode conductor can be made on the load side of the service disconnecting means or the transformer distribution panelboard.

Supply Equipment

Supply equipment must be located on the left side of the parked vehicle, between 8 and 10 feet away from the centerline of the stand, and longitudinally, anywhere from the back end of the stand to a point 15 feet forward from the back of the stand.

A disconnecting means or circuit supply switch must be provided in the source of supply mentioned above.

All supply equipment must be accessible, having a passageway open to it that is at least 2 feet wide and 6.5 feet high.

Supply equipment must be mounted so that the bottom of the enclosure is at least 2 feet above ground level, and the operating means is no more than 6.5 feet above grade.

All equipment must be grounded by a continuous grounding conductor. Receptacles must be arranged so that their removal won't disturb the grounding system.

All outdoor equipment must be rain-tight.

Meter sockets without meters must be enclosed with approved blanking plates.

Overhead conductors (600 volts or less) must have a vertical clearance of 18 feet and a horizontal clearance of 3 feet from all areas subject to recreational vehicle movement.

Underground Service, Feeder, and Branch-Circuit Conductors

All such conductors must be insulated and suitable for such use.

All splices or taps must be made in junction boxes, or if otherwise, by methods approved for such use.

When directly buried conductors leave a trench, they must be protected by one of the following means:

Rigid metal conduit.

Intermediate metal conduit.

Rigid nonmetallic conduit.

Electrical metallic tubing, with supplemental corrosion protection such as bitumastic paint.

Other approved raceways.

The abovementioned protection must extend from grade level down to 18 inches below grade.

Chapter 32

Data Processing Areas

Information Technology Equipment (*Article 645*)

Application

These requirements apply to areas that have all of the following characteristics:

A disconnecting means is installed, which disconnects all power in the area.

The area has its own dedicated heating, air-conditioning, and ventilating system.

Listed electronic and computer data processing equipment is installed in the area.

The only persons permitted in these areas are those who are required for the operation of the equipment installed in them.

This area must be separated from other occupancies by fire-resistant floors, walls, and floors with protected openings. Any openings in firewalls must be properly sealed.

The construction of the building must be in accordance with building codes.

Supply Circuits and Cables

All branch circuits supplying data processing equipment must have an ampacity equal to 125% of the load they serve.

Data processing equipment can be connected to power circuits by any of the following means that are listed for the specific application:

Computer/data processing cables with attachment plugs.

Flexible cords with attachment plugs.

Cord-set assemblies. If these are installed on floor surfaces, they must be protected.

Data processing units are allowed to be interconnected with cables or cable assemblies. If these cables are run on the surface of the floor, they must be protected from physical damage.

Power cables, communications cables, connecting cables, interconnecting cables, and receptacles for data processing equipment are allowed under raised floors, as long as the following conditions are met:

The area under the raised floor is accessible, and the floor is of suitable construction.

Ventilation under the floor area is used only for the data processing equipment and the data processing area.

All openings in the raised floors must protect cables passing through them from damage, and must minimize the amount of debris that can pass through them into the underfloor area.

The branch circuits to receptacles or equipment must be in one of the following:

Rigid metal conduit.

Intermediate metal conduit.

Electrical metallic tubing.

Metal wireway.

Surface metal raceway that has a metal cover.

Flexible metal conduit.

Flexible liquid-tight metal or nonmetallic conduit.

Type MC, MI, or AC cables.

Power cables, communications cables, connecting cables, interconnecting cables, and associated boxes, connectors, plugs, and receptacles for data processing equipment are *not* required to be secured in place if used under raised floors.

Any cables that extend beyond computer rooms are subject to other sections of this code.

Power

There must be one disconnecting means in the area that will disconnect all data processing equipment. There must also be a similar means to turn off all HVAC equipment in the area, and to close all fire and smoke dampers.

The disconnecting means mentioned above must be grouped together at the main exit door(s). A single disconnecting means performing both functions (disconnecting all of the data processing equipment and all of the HVAC equipment, and closing all fire and smoke dampers) may be used rather than two separate (though grouped together) disconnecting means. Integrated electrical systems (as described in *Article 685* of the NEC) are excepted from this requirement.

UPS systems, including their supply and output circuits and battery banks, must all be disconnected from the equipment they serve when the disconnecting means installed at the door is pulled. Integrated electrical systems (as described in *Article 685*) are excepted from this requirement.

Grounding

All data processing equipment must either be grounded (as specified in *Article 250*) or double-insulated.

Power systems that are derived inside data processing equipment (such as in power supplies with transformers) are not considered to be separately-derived systems, in regard to grounding requirements for separately-derived systems (see *Section 250.5[D]*).

All exposed, non–current-carrying metal parts of data processing equipment must be grounded.

Chapter 33

Swimming Pools

Swimming Pools (*Article 680*)

Power Supply and Circuits

Transformers used for supplying power to underwater fixtures and their cases must be identified as suitable for the installation. They must have a grounded barrier between the primary and secondary sides, and must be of the two-winding type.

Either circuit-breaker or receptacle-type ground-fault circuit interrupters can be used.

Conductors from the load side of GFIs (ground-fault interrupters) can't occupy the same enclosures or raceways with conductors that are not GFI-protected. The panels or boxes where the GFI-protected circuits originate are necessarily excepted from this requirement.

All receptacles in a pool area must be at least 10 feet from inside pool walls, except receptacles for necessary pump motors for a permanently-installed pool, which must be at least 5 feet from the inside pool wall and must be of the locking, grounding, and GFI type (Figure 33.1).

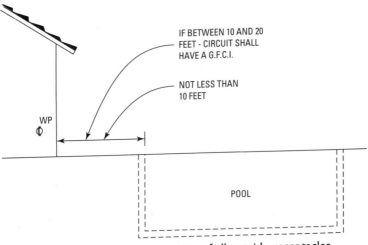

IF BETWEEN 10 AND 20 FEET - CIRCUIT SHALL HAVE A G.F.C.I.

NOT LESS THAN 10 FEET

WP

POOL

Figure 33.1 Location and protection of all outside receptacles.

All permanently installed pools must have at least one GFI-protected receptacle between 10 and 20 feet from an inside pool wall.

All receptacles within 20 feet of the inside pool wall *must* be GFI-protected.

No lighting fixtures can be installed over a pool, or within 5 feet horizontally from the inside pool wall, unless they are at least 12 feet above the highest water level.

There are two exceptions to the preceding rule:

1. If the fixtures are within 5 feet of the inside pool wall, but not above the pool; at least 5 feet above the maximum water level; and are rigidly attached to the existing structure.

2. If the fixtures are totally enclosed, GFI-protected, and at least 7.5 feet above the maximum water level.

Lighting fixtures and outlets that are between 5 and 10 feet from the inside pool wall must be GFI-protected, rigidly attached to the structure, and at least 5 feet above the highest water level.

Cord-connected lighting fixtures, when installed within 16 feet of the surface of the water at any point, must have cords that are no more than 3 feet long and must have a copper equipment-grounding conductor no smaller than No. 12 AWG copper. They must also have an appropriate attachment plug.

No switches or switching devices can be installed closer than 5 feet from the inside walls of a pool, unless separated by a solid fence, wall, or other barrier.

Fixed or stationary equipment rated 20 amperes or less can be connected by the cord-and-plug method. The cord can be no more than 3 feet long and must have a copper equipment-grounding conductor no smaller than No. 12 AWG copper. It must also have an appropriate attachment plug.

Pools can't be located under any overhead wiring (including service drops). No such wiring can be installed above a pool area, which includes any area within 10 feet horizontally measured from an inside pool wall; above a diving area; or above observation stands, platforms, or towers. (Utility-owned, operated, and maintained overhead lines can be installed above such areas. See *Figure 680.1.*) Utility-owned, operated, and maintained overhead communication lines can be installed above such areas if they are at least 10 feet above pools, diving areas, observation stands, platforms, or towers.

All branch circuits supplying electric pool water heaters must have an ampacity equal to 125% of the load they serve.

Underground wiring can't be installed under a pool, or within 5 feet of an inside pool wall except:

1. Wiring necessary to supply pool equipment can be installed in this area.

2. If space considerations demand that wiring be installed within 5 feet from the inside wall of a pool, the following methods can be used:

   Rigid metal conduit that has corrosion protection can be buried at least 6 inches below grade.

   Intermediate metal conduit that has corrosion protection can be buried at least 6 inches below grade.

   Rigid nonmetallic conduit or other approved raceways (all metal raceways require corrosion protection) can be buried at least 18 inches below grade.

Underwater Lighting

All circuits supplying underwater lighting fixtures that operate at more than 15 volts must be GFI-protected.

All underwater lighting fixtures must be approved for the exact circumstances under which they are to be installed. This includes the rating and characteristics of branch circuits that supply the fixtures, as well as the mounting locations.

Bonding

The following parts must be bonded together with a solid copper conductor, no smaller than No. 8 AWG, which can be insulated or bare (Figure 33.2):

   All metallic parts of the pool shell, including the reinforcing metal of the pool shell, deck, and coping stones.

   All forming shells. These form the concrete around niche-type underwater lighting fixtures.

   Metal fittings in or attached to the pool structure.

   All metal parts of electric equipment associated with the pool water system, including pump motors.

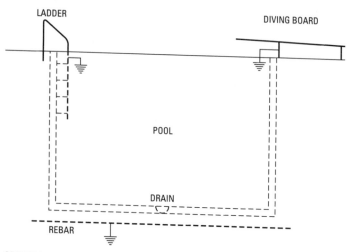

GROUND METAL PARTS, REBAR, LADDER, RAILS,
DIVING BOARDS, DRAINS, OTHER PARTS & EQUIPMENT.

Figure 33.2 Bonding all metal equipment together with solid copper wire.

Metal parts of equipment associated with pool cover systems, including motors.

Metal-sheathed cables, metal raceways, metal piping, and all metal parts that are 5 feet or less from an inside pool wall, 12 feet or less above the maximum water level, or any observation stands or platforms that are not divided from the pool by permanent barriers.

Exceptions to the above requirement are as follows:

1. Reinforcing steel must be bonded together with steel tie wire only.

2. Isolated metal parts that don't measure more than 4 inches in any dimension and don't penetrate more than 1 inch into the pool structure don't have to be bonded.

3. Structural reinforcing steel or the walls of bolted or welded metal pool structures are allowed as a bonding grid for non-electric parts if proper connections are made (see *Section 250-113*).

A bonding grid can be made of any of the following:

The wall of a bolted or welded pool.

A solid copper conductor, no smaller than No. 8 AWG, insulated, bare, or covered.

The reinforcing steel of a concrete pool, if the reinforcing rods are bonded together by steel tie wires or the equivalent.

Some pool water heaters rated more than 50 amps have special bonding requirements, and the instructions with the heaters should be followed.

Underwater Audio Equipment

All underwater audio equipment must be listed for the intended installation, and must be installed according to the manufacturer's instructions.

The following wiring methods can be used for underwater audio equipment:

Rigid metal conduit made of brass or other corrosion-resistant material.

Intermediate metal conduit made of brass or other corrosion-resistant material.

Rigid nonmetallic conduit.

When nonmetallic conduit is used, a No. 8 copper grounding conductor must be run in the conduit and must terminate in the forming shell and in the junction box. In the forming shell, the No. 8 conductor must be covered with a suitable potting compound to prevent deterioration.

All electrical equipment associated with underwater audio equipment must be grounded.

Grounding Requirements

Wet-niche lighting fixtures must have a No. 12 or larger copper grounding conductor in the conduit that feeds them (Figure 33.3).

For wet-niche fixtures that are wired with cords, a separate grounding conductor must be included in the cord. This conductor must be at least as large as the supply conductors, and it must never be smaller than No. 16.

All pool-associated motors must be grounded with at least a No. 12 insulated copper equipment-grounding conductor.

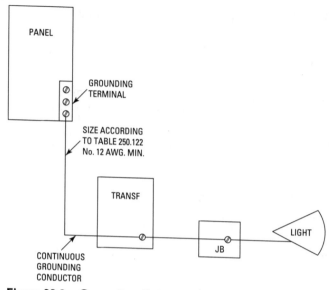

Figure 33.3 Grounding all electrical equipment to the panelboard.

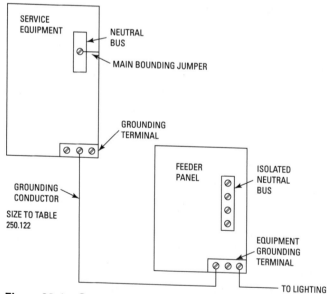

Figure 33.4 Grounding a service panel to the panelboard.

All panelboards must have a separate equipment-grounding conductor to the service panel. This conductor must be sized according to *Table 250.95*, but may never be smaller than No. 12 insulated copper (Figure 33.4).

When cords are used, their equipment-grounding conductors must be connected to a fixed metal part.

Chapter 34

Solar Electric Systems

Solar Photovoltaic Systems (*Article 690*)

Installation

Photovoltaic (solar power) circuits are *not* allowed to use the same raceway, cable, cable tray, outlet or junction box, or similar parts of branch circuits or feeders.

Photovoltaic circuits are permitted to be in the same box if divided from each other by a partition.

Connections to a module or panel must be arranged so that removal of a module or panel won't disconnect a grounded conductor to another photovoltaic circuit.

Photovoltaic arrays mounted on a roof must be provided with ground-fault protection to reduce the risk of fire. This protection must be able to detect a ground fault, interrupt it, and disable the array.

In one- and two-family dwellings, photovoltaic circuits that operate at more than 150 volts to ground may not be accessible to anyone except qualified personnel.

The ratings of conductor ampacities and overcurrent devices must be at least 125% of the calculated current.

All photovoltaic circuits must have overcurrent protection. If a circuit has more than one source of power, overcurrent protection must be provided at each source.

Overcurrent protection must be provided for both sides of a transformer connected to a photovoltaic system. Both sides must be protected as if they were the primary.

Disconnecting Means

A means must be provided to disconnect all photovoltaic circuit conductors from all other conductors. This disconnect must meet the same general requirements as service disconnects (see Chapter 4), but the switches don't have to be service rated, and equipment such as isolating devices, blocking diodes, and overcurrent devices are allowed on the line side of the switch.

The disconnecting means mentioned above must be a manually operated switch or breaker that is readily accessible, is externally operated, indicates whether it is open or closed, and must be rated for the load it disconnects. In cases where both sides of the disconnect can be energized when the switch is open, a warning sign must be installed. The sign must read: "WARNING—ELECTRIC SHOCK—DON'T TOUCH—TERMINALS ENERGIZED IN OPEN POSITION."

Means must also be provided to disconnect all photovoltaic equipment (power conditioners, filters, etc.) from all ungrounded conductors of any sources. If multiple sources energize the equipment, their disconnects must be grouped together and marked.

Fuses that are energized from both directions must be provided with a disconnecting means.

Means must be provided to disconnect all arrays or parts of arrays.

Wiring

All raceway and cable methods outlined in the Code are allowed for photovoltaic installations, as long as they are suitable for the area in which they are installed.

Single-conductor-type UF cables are allowed to be installed in photovoltaic source circuits. Cables exposed to sunlight must be approved for this purpose.

All interconnecting components must have the same insulation and ability to withstand heat as the wiring method used to provide power to the components.

All junction, outlet, and pull boxes located behind panels must be accessible, even if only by the moving of a panel.

Grounding

Two-wire photovoltaic power sources rated over 50 volts must have one wire grounded, and 3-wire photovoltaic power sources must have their neutral conductors grounded.

The grounding connection for dc circuits can be at any one point in the circuit. Locating the grounding point near the source of power will protect the circuit(s) better.

The equipment-grounding conductor must be sized according to the following:

> If the available short-circuit current is less than two times the overcurrent protection rating, a grounding conductor at least the same size as the supply conductors must be used.

> If the available short-circuit current is two or more times the rating of the overcurrent device, the grounding conductor must be sized according to *Section 250.122*.

All exposed metal non–current-carrying parts must be grounded.

Source Connections

Controls must be arranged so that a loss of power from a power-conditioning unit that interacts with other power systems will cause that power unit to be disconnected from all other power sources, and will also ensure that the power-conditioning unit won't be reconnected to other power systems until power from the conditioning is restored.

Normally interactive solar photovoltaic systems are allowed to operate as stand-alone systems to supply wiring in the premises. These systems don't require complex controls, and are allowed to supply power without such controls if they are not interconnected with other power systems.

The power output from an interactive solar photovoltaic system or from a power-conditioning unit must be connected to the supply side of the service-disconnecting means, or connected to the load side as follows:

The connection must be made through a dedicated circuit breaker or fusible disconnect.

The sum of the ratings of overcurrent device currents that feed power to a conductor or bus may not be greater than the ampacity of the bus or conductor, except for systems for a dwelling unit, where this sum can be up to 120% of the conductor or bus's ampacity.

The connection point must be on the line side of all ground-fault protection devices, except that current sources that already have ground-fault protection can be connected to the load side of other ground-fault protection devices.

All back-fed devices must be listed for such use.

In circuits that supply power to conductors or bus bars, any equipment having overcurrent devices must be marked, naming all of the sources of supply to that circuit. This is not required if the circuit has only one supply source.

Storage Batteries

All storage battery installations must conform with the requirements of *Article 480* (see Chapter 22), with the exception that interconnected battery cells are considered grounded when the photovoltaic power source is inherently protected.

In dwellings, the storage batteries must be connected so that the operating voltage is less than 50 volts, except if no parts are accessible, even during maintenance, which is frequently required.

Live parts of battery systems for dwellings must be protected to prevent accidental contact by people or objects.

A means of control (whether inherent or a separate controller) must be used to prevent under- or overcharging of battery installations.

Chapter 35

Emergency Systems

Please note that the many engineering requirements for an emergency system are not mentioned here—only the installation requirements. If you must design part of an emergency system, you must consult the National Electrical Code (NEC) for the design requirements.

Emergency Systems (*Article 700*)

Circuit Wiring

All boxes and enclosures (of all types, including enclosures for equipment such as transfer switches, etc.) must be marked, showing that they are part of an emergency system.

All emergency system wiring must be completely independent from all other wiring systems. It may not share the same raceway, cable, box, or cabinet with other wiring systems.

The above requirement is excepted in the following circumstances:

1. In transfer switch enclosures.

2. In exit or emergency lighting fixtures (or a junction box attached to an exit or emergency fixture) that receive power from two sources.

3. Wiring from two separate emergency systems that receive their power from a common source can share raceways, cables, etc.

4. In a common junction box that contains *only* an emergency circuit and the branch circuit supplying a unit of equipment.

No appliances or lamps (except those that are specifically required for emergency systems) are allowed to be connected to emergency circuits.

Emergency systems must be designed and installed so that when any single lighting element fails (as when a light bulb burns out), no area will be left in total darkness.

When the normal source of lighting in an area is entirely high-intensity discharge lighting (which takes several minutes to get back to original brightness after a power failure)—such as high- or low-pressure sodium, metal halide, or mercury vapor lighting—the emergency lighting system must stay on until the normal lighting system gets back to its full brightness.

This requirement is excepted when other means have been taken to ensure an adequate lighting level.

Branch circuits for emergency lighting must provide power from a second source when normal power is interrupted. This is done by one of the following methods:

> An independent emergency lighting supply, to which the emergency lighting circuit will automatically transfer, in the event of an outage. The emergency lights are connected to the normal source, but are transferred (via an automatic transfer switch) to the emergency power source in case of an outage.
>
> Two or more independent sources of supply, both of which provide enough power for emergency lighting. Unless both systems are kept lighted, a means must be provided for energizing the other upon the failure of one of the systems. Either system (or both) can be a part of the general lighting system if its circuits comply with all other emergency circuit requirements.

Branch circuits that supply emergency classified equipment must be automatically switched to an emergency power source when the normal power fails.

Switches in emergency lighting circuits must be arranged so that they can be controlled only by authorized persons. Usually this is done by using key switches.

The rule above is excepted if two or more single-pole switches connected in parallel are used to control the emergency lighting, and if one of the switches is accessible only to authorized people. Also, accessible switches that can turn emergency lighting on, but not off, are allowed.

No 3-way, 4-way, or series-connected switches are allowed to be used for emergency lighting control.

All manual switches controlling emergency lighting must be accessible to the authorized persons who must operate them. In theaters or places of assembly, at least one switch must be placed in the lobby.

An emergency lighting switch may *never* be placed in a projection room, or on a stage or platform, unless it is merely one of multiple switches and can turn the lighting on, but can't turn it off.

Exterior lights that are not required when daylight is available can be controlled by a photoelectric switch.

Branch-circuit overcurrent devices must be accessible *only* to authorized personnel.

The alternate source of power for emergency situations does *not* have to have ground-fault protection.

Legally Required Standby Systems (*Article 701*)

A sign must be placed at the building entrance, detailing the types and locations of legally-required standby power sources.

Wiring for legally-required standby systems *is* allowed to use the same raceways, cables, boxes, etc. as general wiring systems.

Branch-circuit overcurrent devices must be accessible *only* to authorized personnel.

The alternate source of power for emergency situations does *not* have to have ground-fault protection.

Optional Standby Systems (*Article 702*)

A sign must be placed at the building entrance, detailing the types and locations of optional standby power sources.

Wiring for optional standby systems *is* allowed to use the same raceways, cables, boxes, etc., as general wiring systems.

Chapter 36

High Voltage

Wiring Over 600 Volts (*Article 490*)

Author's Note

Although wiring over 600 volts is usually called "high voltage," this is not technically correct. Voltages between 600 and 35,000 volts are properly called "medium voltage," and voltages over 35,000 volts are properly called "high voltage."

Wiring Methods

In above-ground locations, conductors can be installed as follows:

In rigid metal conduit.

In intermediate metal conduit.

In rigid nonmetallic conduit.

In a cable tray.

As busways.

As cablebus.

In other identified raceways.

As open runs of metal-clad cable, when suitable for the use.

When accessible to qualified personnel only:

As open runs of type-MV cables.

As bare conductors or bus bars.

Bus bars can be either copper or aluminum.

Open runs of insulated wires or cables that have lead sheathing or a braided outer covering must be properly supported to avoid physical damage and electrolytic damage due to the contact of dissimilar metals.

The minimum distances between bare conductors and other conductors or grounding surfaces must be no less than the values shown in *Table 490.24*.

Underground Conductors

Underground conductors must be identified for the specific voltage and conditions for which they are installed.

Directly buried cables over 2000 volts must be shielded, except for nonshielded multiconductor cables between 2001 and 5000 volts that have an overall metallic sheath or armor.

The depths of underground conductors (in raceways or as directly buried cables) must be in accordance with *Table 300.50.*

Nonshielded cables must be installed in one of the following:

> Rigid metal conduit, encased in at least 3 inches of concrete.
>
> Intermediate metal conduit, encased in at least 3 inches of concrete.
>
> Rigid nonmetallic conduit, encased in at least 3 inches of concrete.
>
> Type MC cable, when the metal sheath is grounded by an effective grounding path (see *Section 250.97*).
>
> Lead-sheathed cable, when the lead sheath is grounded by an effective grounding path (see *Section 250*).

Conductors emerging from the ground are required to be protected by an appropriate raceway.

Conductors installed on poles must be in one of the following, from the ground up to 8 feet above grade:

> Rigid metal conduit.
>
> Intermediate metal conduit.
>
> Schedule 80 PVC conduit.
>
> An equivalent protective means.

Conductors entering a building must be protected by an approved enclosure, from the ground line up to the point of entrance.

All metallic enclosures must be grounded.

Directly buried cables may be spliced in a splice box, or without a splice box, as long as the splice is made with an approved method, suitable for the conditions encountered. Such splices must be mechanically protected and watertight. If the cables are shielded, the shield must be continuous across the splice.

Backfill must not contain materials such as large or sharp rocks, which could damage directly buried conductors.

Sand or sleeves may be used to protect the cable(s).

Voltage-stress-reduction means must be provided at all terminations of factory-applied shielding.

Cable sheath terminating devices must be used wherever protection from moisture and physical damage is necessary.

Equipment

Author's Note
Equipment for high voltage installations should be designed not only for the specific use, but also for the specific installation. There is a considerable amount of danger involved with these voltages, and every installation should be thoroughly engineered.

Pipes or ducts that require occasional maintenance, or whose malfunction would endanger the operation of electrical equipment, may not be located in the vicinity of service equipment, switchgear, or controllers. Protection from such problems as leaks must be provided.

Fire protection piping is specifically allowed.

Overcurrent protection must be provided for each conductor by an approved means.

Any doors that would give access by unauthorized people to energized parts *must* be locked.

Control and instrument switches or pushbuttons must be accessible and mounted no higher than 6½ feet (2 meters) above the floor. Operating handles that require more than 50 pounds of force to activate may not be higher than 66 inches from the floor. Operating handles for infrequently used devices may be located anywhere they can be serviced by a mobile platform.

Electrode-Type Boilers
Electrode-type boilers can be supplied *only* with a 3-phase, 4-wire, solidly grounded wye system; or by means of an isolation transformer that derives such a system.

Control circuits for electrode-type boilers must operate at 150 volts or lower, and must be taken from a grounded source. No controls are allowed in a grounded conductor.

Circuits to boilers must be rated at 100% of the total load.

Circuits such as those mentioned above must have 3-pole ground-fault protection. This protection is allowed to automatically reset following an overload, but not following a ground-fault.

Phase-fault protection must be in each phase.

Means must be taken for the detection of neutral and ground currents.

The neutral conductor must:

Be connected to the pressure vessel containing the electrodes.

Be insulated for no less than 600 volts.

Have an ampacity that is no less than that of the largest ungrounded branch-circuit conductor.

Be installed in the same raceway or cable tray as its circuit conductors.

Not be used for any other circuit.

All exposed non–current-carrying metal parts of electrode-type boilers and associated structures and parts must be bonded to the pressure vessel, or to the neutral conductor to which the pressure vessel is attached. The ampacity of the bonding conductor may not be less than that of the neutral conductor.

Each boiler must be equipped with a means to limit its maximum temperature and pressure by interrupting (either directly or indirectly) the current flow to the electrode. This protection must be in addition to any other temperature and pressure regulation and any pressure-relief or safety valves.

Chapter 37

Low Voltage

Circuits and Equipment Operating at Less Than 50 Volts (*Article 720*)

Conductors smaller than No. 12 copper are not allowed.

Conductors that supply more than one receptacle or appliance can't be smaller than No. 10 copper.

All lampholders used with these systems must be rated at least 660 watts.

Receptacles must have a rating of at least 15 amperes.

Receptacles of 20 amps or more are required in kitchens, laundries, or other locations where portable appliances are likely to be used.

Overcurrent protection is required. (See *Article 240* and Chapter 6.)

Circuits less than 50 volts must be grounded under the following conditions:

> When these circuits are supplied by transformers with a supply voltage of over 150 volts.
>
> When these circuits are supplied by transformers with ungrounded supplies.
>
> When these circuits are installed as overhead circuits outside of buildings.

Class 1, 2, or 3 Remote-Control, Signaling, and Power-Limited Circuits (*Article 725*)

The circuits covered by this article are remote-control, signaling, and power-limited circuits that are not an integral part of another device or appliance.

Classifications

A remote-control, signaling, or power-limited circuit is considered to be the portion of the wiring system between the load side of the overcurrent device or power-limited supply and all connected equipment. These circuits must be class 1, class 2, or class 3. The three classes of circuits are as follows:

Class 1 power-limited circuits—The sources that supply these circuits must have a rated output of no more than 30 volts and 1000 volt-amperes (the term *volt-amperes* is abbreviated VA and is essentially the same thing as watts). When power sources other than transformers are used to supply these circuits, the sources must be protected by overcurrent devices that are rated at no more than 167% of the rated current (volt-amps divided by rated voltage). The overcurrent device may be a part of the power supply, and can't be interchangeable with overcurrent devices of a higher rating.

Class 2 and class 3 circuits—These circuits are limited either inherently, or by a combination of power source and overcurrent limitations.

Remote-control circuits that supply safety control equipment must be classified as class 1 if failure of the equipment would cause a fire hazard or other hazard to life. Such equipment as controls for heating or air-conditioning equipment is not to be considered as safety control equipment.

Class 1 circuits can't be run in the same cable as communications circuits. Class 2 and class 3 circuits can be run in the same cable as communications circuits, but when so installed, they must be considered communications circuits and must comply with the requirements of *Article 800* (see Chapter 39). Cables specifically designed to be used in this way can be used, with the class 2 and 3 conductors retaining their class 2 and 3 ratings.

Class I Circuits

The sources that supply these circuits must have a rated output of no more than 30 volts and 1000 volt-amperes. When power sources other than transformers are used to supply these circuits, the sources must be protected by overcurrent devices that are rated at no more than 167% of the rated current (volt-amps divided by rated voltage). The overcurrent device may be a part of the power supply, and can't be interchangeable with overcurrent devices of a higher rating.

When transformers are used to supply these circuits, they must comply with *Article 450*.

Power sources for these circuits other than transformers can have a maximum power output of 2500 VA. Note that the *rated* power was covered above, and in this paragraph the *maximum possible output* is discussed. Also, the maximum possible voltage and the maximum possible current should be multiplied, and the total can be no more than 10,000. These ratings must be calculated based on there being no overcurrent devices in the circuit, which would not occur in a proper installation. This set of requirements allows for a worst-possible-case scenario.

Class 1 remote-control and signaling circuits may operate on up to 600 volts, and the power output of the source does not have to be limited.

Conductors No. 14 or larger must have overcurrent protection. Derating factors can't be applied. No. 18 conductors must be protected at no more than 7 amps, and No. 16 conductors must be protected at no more than 10 amps.

Exceptions to this rule are as follows:

1. Where other parts of the NEC require (or permit) different levels of protection.

2. Class 1 conductors that are supplied by 2-wire transformer secondaries can be supplied with overcurrent protection only on the primary side of the transformer.

3. Class 1 circuit conductors that are No. 14 or larger and tapped from the load side of the overcurrent device of controlled power circuits are only required to have ground-fault and short-circuit protection. If the rating of the branch-circuit protective device is not more than three times the rating of the circuit conductors, no other protection is required.

Overcurrent devices must be located at the point of supply, except as follows:

1. When the overcurrent protection device for a larger conductor also protects a smaller conductor.

2. When class 1 conductors that are supplied by 2-wire transformer secondaries can be supplied with overcurrent protection only on the primary side of the transformer.

Class 1 circuits must be installed according to the requirements of other parts of the Code, except as modified by the requirements of this section.

Class 1 circuits can occupy the same raceways, cables, etc. as other circuits, provided all the conductors are insulated for the maximum voltage of any conductor. Note that in this subsection we are referring to conductors of different circuits, not conductors of different systems.

Power supply circuits and class 1 circuits can be run in the same cable, raceway, etc. only when they are associated, except in control centers or as underground conductors in a manhole, as long as one of the following conditions is met:

> The conductors are in a metal-enclosed or UF cable.
>
> The conductors are separated from the power supply conductors by a fixed nonconductor (such as plastic tubing) in addition to the insulation of the conductor.
>
> The conductors are separated from the power supply conductors by mounting on racks, etc.

Conductors of No. 16 or No. 18 can be used, as long as they don't exceed the ampacities given in *Table 402.3*. They must be installed in a listed raceway, enclosure, or cable. Conductors larger than No. 16 must comply with *Section 310.15*, and flexible cords must comply with *Article 400*.

Insulation must be rated at least 600 volts. Wires larger than No. 16 must comply with *Article 310*, and wires No. 16 or No. 18 must have one of the following types of insulation: RFH-2, RFHH-2, RFHH-3, FFH-2, TF, TFF, TFN, TFFN, PF, PFF, PGF, PGFF, PTF, PTFF, SF-2, SFF-2, PAF, PAFF, ZF, ZFF, KF-2, KFF-2, or another type listed for such use.

When only class 1 conductors are in a raceway, the allowable number of conductors can be no more than easy installation and the dissipation of heat allow.

Where power supply and class 1 conductors are allowed in the same can apply only as follows:

> To all conductors, when the class 1 conductors carry continuous loads exceeding 10% of their rated load, and where there are more than three conductors.
>
> To only the power supply conductors, when there are more than three such conductors, and when the class 1 conductors don't have continuous loads of more than 10% of their rated current.
>
> Class 1 circuit conductors can be installed in cable trays, according to the requirements of tray cables. (See *Sections 392.8* through *392.11*.)
>
> Where damage is likely, all conductors must be installed in rigid metal conduit, intermediate metal conduit, rigid nonmetallic conduit, EMT, MI cable, or MC cable, or be protected from damage by some other effective means.

Class 2 and Class 3 Circuits

The power for class 2 or class 3 circuits must be limited either inherently, or by a combination of power source and overcurrent limitations. (See *Figure 725.41* of the NEC.)

These power supplies can't be paralleled or interconnected unless listed for such use.

When overcurrent protection is required, the devices can't be interchangeable with devices of higher ratings. The device may be part of the power supply.

These overcurrent devices must be installed at the point of supply.

Wiring methods on the supply side must comply with all other requirements of the NEC. Transformers and like devices that are supplied by light or power circuits must be protected at no more than 20 amps, except that input leads of transformers can be as small as No. 18 if they are no more than 12 inches long.

Wiring materials and methods on the load side must be listed for the purpose and must be installed according to the following guidelines:

> Class 2 or 3 conductors must be separated at least 2 inches from the other conductors, except as follows:
>
> 1. When either the light, power, or class 1 conductors, or class 2 or 3 conductors, are enclosed in raceway or metal-sheathed, metal-clad, nonmetallic-sheathed, or UF cables.
>
> 2. Where the conductors of the differing systems are separated by a fixed insulator in addition to the insulation of the conductors, such as a porcelain or plastic tube.

> Class 2 or 3 conductors can't be installed in any cables, cable trays, enclosures, or raceways with power, lighting, or class 1 circuits, except as follows:
>
> 1. Where they are separated by a barrier.
>
> 2. In enclosures, etc. when class 1 conductors enter only to connect equipment to which the class 2 and 3 circuits are connected also.
>
> 3. Underground conductors in manholes, when the power or class 1 conductors are in UF cable or metal-enclosed cable *or* the conductors are separated by a fixed insulator, in addition to the insulation of the conductors *or* the conductors are firmly mounted on racks, etc. *or* when installed as part of a hybrid cable for a closed-loop system. (See *Article 780.*)

In hoistways, class 2 or 3 conductors must be installed in rigid
metal conduit, rigid nonmetallic conduit, intermediate metal
conduit, or EMT, except as allowed in elevators. (See *Section
620.21.*)

In shafts, class 2 and 3 conductors must be separated at least
2 inches from other systems.

Two or more class 2 circuits can be installed in the same cable,
enclosure, or raceway, as long as every conductor is insulated sufficiently for the highest voltage of any conductor present.

Conductors of two or more class 3 circuits are allowed to share
the same raceway, cable, etc.

Class 2 circuits can share the same raceway, cable, or enclosure
with class 3 circuits, as long as the insulation of the class 2 conductors is equivalent to the insulation required for the class 3 conductors.

Class 2 or 3 circuits are allowed to be in the same raceway or
enclosure with other circuits, provided they are in a jacketed cable
of one of the following types:

Power-limited fire alarm circuits, according to *Article 760.*

Optical fiber cables, according to *Article 770.*

Communications circuits, according to *Article 800.*

Community antenna television systems and radio distribution
systems, according to *Article 820.*

When these conductors are extended out of a building and are
subject to accidental contact with other systems that operate at
over 300 volts to ground, they must meet the requirements for communications circuits. (See *Article 800* and Chapter 40.)

All cables used for class 2 and 3 systems must be listed as resistant to the spread of flames.

Class 2 and 3 cables are marked, listing the areas they are
designed for. All class 2 and 3 cables must be installed *only* in areas
they are listed for.

Abandoned cables must be removed. Exceptions are made for
cables that are not accessible.

Chapter 38

Fiber-Optic Cables

Optical Fiber Cables (*Article 770*)

Brief Explanation

Optical fiber cables (also called fiber-optic cables) are included in the National Electrical Code (NEC) for two primary reasons:

Because they are usually installed by the same persons who install electrical wiring.

Because optical fiber systems interact with, and depend upon, electrical and electronic systems.

The inclusion of optical fiber cables in the NEC is unusual, in that these cables carry no electricity at all. Nevertheless, they connect to electronic equipment and are usually installed by the same people who install electrical and electronic wiring. And, of course, optical fibers are contained within cables, which need proper fire ratings.

These cables have three basic classifications, which have to do with the type of cable jacketing and construction, rather than with the operation of the optical fibers that are contained in the cables. The classifications are as follows:

Horizontal cables.

Riser cables.

Plenum cables.

These classifications are subdivided into two divisions each:

Conductive.

Nonconductive.

Conductive cables have non–current-carrying metal members in the cables. These cables are used as strength members, metal sheaths, or vapor barriers.

Nonconductive cables are cables without any such metal members.

One other type of cable is called a *hybrid* cable. These cables contain not only optical fibers, but also current-carrying conductors and possibly non–current-carrying conductive members as well.

Installation

Where optical fiber cables with non–current-carrying conductive members contact electrical power or light conductors, the conductive member must be grounded as close as possible to the point of entrance. In place of grounding, an insulating joint can be installed in the conductive member, as close as possible to the point of entrance.

The point of entrance mentioned above is the place where the fiber-optic cable emerges through an exterior wall, concrete floor slab, or from a grounded rigid or intermediate metal conduit.

Nonconductive optical fiber cables can occupy the same cable tray or raceway as electrical power, light, or class 1 conductors operating at no more than 600 volts.

Hybrid optical fiber cables containing only power, light, or class 1 conductors operating at no more than 600 volts may be installed in the same cable tray or raceway as electrical power, light, or class 1 conductors operating at no more than 600 volts.

Nonconductive optical fiber cables *can't* occupy the same box, panel enclosure, etc., as the terminations of electric light, power, or class 1 circuits, except as follows:

1. When the nonconductive optical fiber is functionally associated with the power, light, or class 1 circuits.

2. When the nonconductive optical fibers are installed in a factory-assembled or field-assembled control center.

3. Nonconductive optical fiber cables are allowed with circuits operating at 600 volts or less in industrial buildings, when supervised only by qualified persons.

4. Hybrid optical fiber cables are allowed with circuits operating at 600 volts or less in industrial buildings, when supervised only by qualified persons.

Conductive and nonconductive fiber-optic cables are allowed in the same raceway, cable tray, or enclosure with any of the following:

Class 2 and 3 remote-control, signaling, and power-limited circuits.

Power-limited fire protection signaling systems.

Communications circuits.

Community antenna television and radio distribution systems.

Abandoned cables must be removed. Exceptions are made for cables that are not accessible.

In plenums, the following types of cables are allowed to be installed:

Type OFNP. This type of cable can also be installed in optical fiber raceways.

Type OFCP.

Other types, installed in approved raceways (see *Section 300.22* and Chapter 1).

For risers, the following types of cables are allowed to be installed:

Type OFNR.

Type OFCR.

Type OFNP.

Types OFN or OFC can be installed in one- and two-family dwellings, or in other occupancies if installed in a metal raceway or fireproof shaft that has firestops installed at each end.

Cables in other locations can be types OFC and OCN.

Types and Uses

Optical fiber cables must be installed according to their listings. The listing designations are as follows:

OFNP—nonconductive optical fiber plenum cable.

OFCP—conductive optical fiber plenum cable.

OFNR—nonconductive optical fiber riser cable.

OFCR—conductive optical fiber riser cable.

OFN—nonconductive optical fiber cable.

OFC—conductive optical fiber cable.

Types OFNP and OFCP can be used in ducts, plenums, and other spaces used for environmental air. They have low-smoke and fire-resistant characteristics.

Types OFNR and OFCR can be used in a vertical shaft that runs from floor to floor. They have fire-resistant characteristics that prevent the spread of fire from floor to floor.

Types OFN and OFC can be used for general installations, but not in plenums, etc. or as risers. They are also resistant to the spread of fire.

Hybrid cables must be used exactly as listed and marked on the cable jacket.

The following types of cables can be substituted for each other:

For OFC—OFN, OFCR, OFNR, ORCP, OFNP.

For OFN—OFNR, OFNP.

For OFCR—OFNR, OFCP, OFNP.

For OFNR—OFNP.

For OFCP—OFNP.

For OFNP—None.

Chapter 39

Communications

Communications Systems (*Article 800*)

Conductors Entering Buildings

If communications and power conductors are supported by the same pole, or run parallel in span, the following conditions must be met:

> Unless necessary otherwise, communications conductors should be located below power conductors.
>
> Communications conductors can't be connected to cross-arms.
>
> Power service drops must be separated from communications service drops by at least 12 inches.

Above roofs, communications conductors must have the following clearances:

> Flat roofs—8 feet.
>
> Garages and other auxiliary buildings—none required.
>
> Overhangs, where no more than 4 feet of communications cable will run over the area—18 inches.
>
> Where the roof slope is 4 inches rise for every 12 inches horizontally—3 feet.

Underground communications conductors must be separated from power conductors in manholes or handholes by brick, concrete, or tile partitions.

Communications conductors should be kept at least 6 feet away from lightning protection system conductors.

Protection

Protectors are essentially surge arrestor devices designed for the specific requirements of communications circuits.

Protectors must be installed on all aerial circuits that are not confined within a block (i.e., city block).

Protectors must also be installed on all circuits within a block that could accidentally contact power circuits over 300 volts to ground.

All protectors must be listed for the type of installation.

Metal sheaths of any communications cables must be grounded or interrupted with an insulating joint as close as practical to the point where they enter a building.

The point of entrance mentioned above is the place where the communications cable emerges through an exterior wall or concrete floor slab, or from a grounded rigid or intermediate metal conduit.

Grounding conductors for communications circuits must be copper or some other corrosion-resistant material, and have insulation suitable for the area of installation.

Communications ground conductors may be no smaller than No. 14.

The grounding conductor must be run as directly as possible to the grounding electrode, and must be protected if necessary.

If the grounding conductor is protected by metal raceway, it must be bonded to the grounding conductor on both ends.

The grounding electrode for communications grounds may be any of the following:

1. The grounding electrode of an electrical power system.
2. A grounded interior metal piping system. Avoid gas piping systems for obvious reasons.
3. Metal power service raceways.
4. Power service equipment enclosures.
5. A separate grounding electrode.

If the building being served has no grounding electrode system, the following can be used as a grounding electrode:

Any acceptable power system grounding electrode or grounded metal structure.

A ground rod or pipe at least 5 feet long and ½ inch in diameter. This rod should be driven into damp (if possible) earth and kept separate from any lightning protection system grounds or conductors. Steam or hot water pipes can't be used; neither may gas pipes be used.

Connections to grounding electrodes must be made via approved means.

If the power and communications systems use separate grounding electrodes, they must be bonded together with a No. 6 copper conductor. Other electrodes may be bonded also. This is not required for mobile homes.

For mobile homes, if there is no service equipment or disconnect within 30 feet of the mobile home wall, the communications circuit must have its own grounding electrode. In this case, or if the mobile home is connected with cord-and-plug, the communications circuit protector must be bonded to the mobile home frame or grounding terminal with a copper conductor no smaller than No. 12.

Communications Conductors in Buildings

Communications conductors must be kept at least 2 inches away from power or class 1 conductors, unless they are permanently separated from these other conductors or unless the power or class 1 conductors are enclosed in one of the following:

Raceway.

Cable of type AC, MC, UF, NM, or NMB, or metal-sheathed cable.

Communication cables are allowed in the same raceway, box, or cable with any of the following:

Class 2 and 3 remote-control, signaling, and power-limited circuits.

Power-limited fire protection signaling systems.

Conductive or nonconductive optical fiber cables.

Community antenna television and radio distribution system.

Communications conductors are not allowed to be in the same raceway or fitting with power or class 1 circuits.

Communications conductors are not allowed to be supported by raceways, unless the raceway runs directly to the piece of equipment the communications circuit serves.

Openings through fire-resistive floors, walls, etc. must be sealed with an appropriate fire-stopping material.

Any communications cables used in plenum or environmental air-handling spaces must be listed for such use.

Communications and multipurpose cables can be installed in cable trays.

Any communications cables used in risers must be listed for such use.

Abandoned cables must be removed. Exceptions are made for cables that are not accessible.

Cable substitution types are shown in *Table 800.53*.

Chapter 40

Special Installations

In addition to all common types of wiring and electrical systems, the National Electrical Code covers a number of very unusual or special types of installations. All of these installation types require careful engineering and detailed installation instructions. Often, the installations vary so greatly that an installation in one location bears little resemblance to the same type of installation in a different location.

Brief descriptions of the special installations covered in the National Electrical Code (and one not covered) follow.

Lightning Protection Systems (Not in NEC)

Lightning protection systems are often installed by electricians but are not covered in the NEC. They are, however, covered in the standard *NFPA 78*, a short book that includes all of the installation requirements. This book should be read before any lightning protection installation is attempted.

Closed-Loop and Programmed Power Distribution (*Article 780*)

This article includes requirements for a new type of electrical system originally designed for the National Association of Home Builders' new Smart House™ system. This system, which has certain advantages over existing power systems, has not yet been widely used.

Radio and Television Equipment (*Article 810*)

This article covers the requirements for television and radio transmitting and receiving equipment.

Community Antenna TV and Radio Distribution Systems (*Article 820*)

This article covers the requirements for coaxial signal distribution for community-owned central antenna systems. This article also covers cable television wiring in private facilities, but not the cable TV wiring owned by utilities. One important requirement of this article is that abandoned coaxial cables must be removed. Exceptions are made for cables that are not accessible.

Fire Alarm Systems (*Article 760*)

This article covers the requirements (similar to those for class 1 and 2 circuits) for fire alarm and similar systems.

Interconnected Electrical Power Sources (*Article 705*)

This article covers the requirements for connecting power sources in parallel.

Irrigation Machines (*Article 675*)

The title is self-explanatory.

Industrial Machinery (*Article 670*)

The title is self-explanatory.

Electroplating (*Article 669*)

The title is self-explanatory.

Electrolytic Cells (*Article 668*)

The title is self-explanatory.

Pipe Organs (*Article 650*)

The title is self-explanatory.

X-Ray Equipment (*Article 660*)

The title is self-explanatory.

Induction Heating Equipment (*Article 665*)

The title is self-explanatory.

Sound Recording Equipment (*Article 640*)

The title is self-explanatory.

Electric Welders (*Article 630*)

The title is self-explanatory.

Elevators (*Article 620*)

The title is self-explanatory.

Marinas and Boatyards (*Article 555*)

The title is self-explanatory.

Agricultural Buildings (*Article 547*)

The title is self-explanatory.

Floating Buildings (*Article 553*)

The title is self-explanatory.

Glossary

Accessible—Can be removed or exposed without damaging the building structure. Not permanently closed in by the structure or finish of the building.

Accessible, readily—Can be reached quickly, without the need to climb over obstacles or use ladders.

Aggregate—A masonry substance that is poured into place, and then sets and hardens, as concrete.

Ampacity—The amount of current (measured in amperes) that a conductor can carry without overheating.

Ampere (or amp)—Unit of current measurement. The amount of current that will flow through a 1-ohm resistor when 1 volt is applied.

Appliance—Equipment that is installed or connected as a unit, to perform functions such as clothes or dish washing, food mixing, etc.

Approved—Acceptable to the authority that has jurisdiction.

Automatic—Self-acting. Operating by its own mechanism, based on a nonpersonal stimulus.

Bonding—The permanent joining of metal parts to form an electrically conductive path.

Branch circuit—Conductors between the last overcurrent device and the outlets.

Branch circuit, appliance—A branch circuit that supplies current to one or more outlets that serve appliances.

Branch circuit, general-purpose—A branch circuit that supplies outlets for lighting and power.

Branch circuit, individual—A branch circuit that supplies only one piece of equipment.

Branch circuit, multi-wire—A branch circuit having two or more ungrounded circuit conductors, with a voltage difference between them, and a grounded circuit conductor (neutral) having an equal voltage difference between it and each ungrounded conductor.

Building—A structure that is either standing alone or cut off from other structures by firewalls.

Cabinet—A flush or surface enclosure with a frame, mat, or trim, on which are mounted swinging doors.

Concealed—Made inaccessible by the structure or finish of the building.

Conduit body—The part of a conduit system, at the junction of two or more sections of the system, that allows access through a removable cover. Most commonly known as condulets, LBs, LLs, LRs, etc.

Continuous load—A load whose maximum current continues for three hours or more.

Controller—A device or group of devices that control (in a predetermined way) power to a piece of equipment.

Cross-sectional area—The area (in square inches or circular mils) that would be exposed by cutting a cross-section of the material.

Cutout box—A surface mounting enclosure with a hinged door.

Device (also used as wiring device)—The part of an electrical system that is designed to carry, but not use, electrical energy.

Disconnecting means—A device that disconnects a group of conductors from their source of supply.

Enclosed—Surrounded by a case, housing, fence, or walls that prevent unauthorized people from contacting the equipment.

Exposed—Able to be inadvertently touched or approached.

Feeder—Circuit conductors between the service and the final branch-circuit overcurrent device.

Ground—An electrical connection (intentional or accidental) between an item of equipment and the earth.

Hoistway—A shaftway or other vertical opening or space through which an elevator or dumbwaiter operates.

Identified (for use)—Recognized as suitable for a certain purpose, usually by an independent agency, such as UL.

Isolated—Not accessible unless special means of access are used.

Lamp—A light source. Reference is to a light bulb rather than a table lamp.

Location, damp—A partially protected location, such as under a canopy, roofed open porch, etc. Also, an interior location that is subject only to a moderate degree of moisture, such as a basement, barn, etc.

Location, dry—An area that is not normally subject to water or dampness.

Location, wet—A location underground, in a concrete slab, where saturation occurs, or outdoors.

Outlet—The place in the wiring system where the current is taken to supply equipment.

Overcurrent—Too much current.

Pendant—A hanging electrical cord, to which is attached a lampholder or receptacle.

Phase converter—A device that derives 3-phase power from single-phase power. Used extensively to run 3-phase equipment in areas (often rural areas) where only single-phase power is available.

Photovoltaic—Changing light into electricity.

Plenum—A chamber that forms part of a building's air distribution system, to which one or more ducts connect. Frequently, areas over suspended ceilings or under raised floors are used as plenums.

Radius (radii, pl.)—The distance from the center of a circle to its outer edge.

Rotary convertor—A type of phase convertor. (See **phase convertor.**)

Separately derived system—A system whose power is derived from a generator, transformer, or convertor.

Service—Equipment and conductors that bring electricity from the supply system to the wiring system of the building being served.

Service drop—Overhead conductors from the last pole to the building being served.

Utilization equipment—Equipment that uses electricity.

Whip—A flexible assembly, usually of THHN conductors in flexible metal conduit with fittings, generally used to bring power from a lighting outlet to a lighting fixture.

Index